A PRACTICAL INTRODUCTION TO

# Digital Command Control for Railway Modellers

KATO
PRECISION RAILROAD MODELS
N Scale
GE AC4400CW
KATO
PRECISION RAILROAD MODELS
N Scale
EMD SD70MAC
GAUGEMASTER®
DCC01
PRODIGY ADVANCE
CONTROL SOLUTIONS
GAUGEMASTER
Controllers
• models for all scales from Z to G
• available in cased, panel mount & handheld formats
• robust build quality
• designed for precision running
DT400
Digitrax
Standard DCC
Easy to use
Simple programming
ZTC-511 Digital Master Controller
ZTC CONTROLS
CLEAR
LOCO
CONTROL
FUNCTION
REMOTE
PROGRAM
AUTO
OFF
ON
BRAKE
FWD
MID
REV
DIGITAL
D.C
OPEN
LH 100
Esc
Enter
DIGITAL
plus
STOP
Roco
47371

A PRACTICAL INTRODUCTION TO

# Digital Command Control for Railway Modellers

NIGEL BURKIN

THE CROWOOD PRESS

First published in 2008 by
The Crowood Press Ltd
Ramsbury, Marlborough
Wiltshire SN8 2HR

**www.crowood.com**



**British Library Cataloguing-in-Publication Data**
A catalogue record for this book is available from the British Library.

ISBN 978 1 84797 020 6

Typeset by Carolyn Griffiths, Cambridge

Printed and bound in Malaysia by Times Offset (M) Sdn Bhd

# Contents

## ABBREVIATIONS

| | |
|---|---|
| ac | Alternating current. |
| AWG | American Wire Gauge. |
| BNSF | Burlington Northern Santa Fe Railway – a US Class 1 railroad company. |
| CTC | Centralized Traffic Control – a US-based signalling system. |
| CV | Configuration Variable (decoder programming slot). |
| dc | Direct current. |
| DCC | Digital Command Control. |
| DPDT | Double pole, double throw switch. |
| DRS | Direct Rail Services. |
| EMD | Electromotive Division – builder of the European-operated Class 66 and locomotives used in the USA. |
| EWS | English, Welsh and Scottish Railway. |
| F1 to F6 etc. | F stands for 'function' (there may be as many as twenty-eight functions that can be controlled by a throttle). References to function button numbers may be prefixed with the letter F in DCC instruction manuals. |
| GBRf | GB Railfreight, a First Group company. |
| GE | General Electric – a US-based locomotive builder. |
| HEX | Hexadecimal. |
| LCD | Liquid crystal display. |
| LED | Light emitting diode. |
| MOROP | Union Européenne des Modélistes Ferroviaires et des Amis des Chemins de Fer (European Union of Model Railroad and Railroad Fans). |
| MRC | Model Rectifier Corporation. |
| MRL | Montana Rail Link. |
| NEM | des NEM du MOROP – NEM standards. |
| NMRA | National Model Railway Association. |
| PCB | Printed circuit board or copper laminated circuit board. |
| RPs | Recommended Practices. |
| RTR | Ready-to-run. |
| SMT LED | Surface-mount LED. |
| TOPS | Total Operations Processing System. |
| USP | Uninterruptible Signal Processing. |
| ZTC | ZTC Controls – a UK-based DCC equipment manufacturer. |

## DEDICATION

This book is dedicated to my ever loving and supportive wife, Sarah.


## ACKNOWLEDGEMENTS

I am indebted to the following for their kind assistance with the production of this book. They have helped with practical advice, proofing for technical content or have contributed equipment featured in the projects.

Alan Murray of MacKay Models.
Stanley Kerr of MacKay Models.
Dennis Lovett of Bachmann Europe Plc.
John Jewitt of Sunningwell Command Control.
David Nicholson of ZTC Controls.
Charlotte Nicholson of ZTC Controls.
Robin Hamp of Express Models.
Ian Fowler of Gaugemaster Controls.
Diana Johnson of Tumuli Design for illustrations and photography.

A special acknowledgement to Pete Lawrenson, Chief of Security and Safety, together with Ben Hampson of Montana Rail Link for their kind assistance with my research into US locomotive operations.

And finally, to Steve Jones of Telford, Shropshire, a well-known DCC enthusiast, for dropping me into all of this DCC stuff in the first place!

# Introduction

We're all beginners at something at some stage in our lives and Digital Command Control (DCC) remains something of a mystery for some modellers (including myself at one time). Although there are many that would like to take the plunge, 'technology' and 'cost' can be barriers to adoption, even though it is becoming generally accepted as an exciting technology that brings added realism to the operation of a model railway. In this book I offer a practical introduction to DCC, accompanied by a variety of projects that can be followed and adapted to suit almost any situation.

There is little doubt that DCC is growing in popularity in the UK and the introduction of accessible entry-level and mid-range sets from Lenz, Bachmann, Gaugemaster and Hornby will help the modeller teetering on the cusp of transition to make the change. I was a convert to DCC in 2001 following a couple of years of my own brand of indecision: humming and hawing over the cost, the perception that DCC is complex and what the potential benefits may or may not be. Having made the decision to 'go digital' after trying it on a friend's loft layout, in July 2001 I purchased a Digital plus Compact system by Lenz Elektronik GmbH with a handful of decoders. It was a major purchase for me and one that was very carefully considered. Five years on and I wonder what the fuss was all about!

The rationale behind my decision to 'go digital' needs to be explained. The construction of the first of my two (then) new exhibition layouts, Dudley Heath Yard, was followed by an intense session of wiring and control panel construction to equip it with 'cab control' to work with a pair of KPC hand-held controllers. The layout was a compact end-to-end design with a scenic section of 8ft supported by 4ft of fiddle yard. To obtain the operational flexibility for shunting and train operation, the layout was divided into nine sections controlled through double pole, double throw (DPDT) switches to enable dual control. In addition, there were six isolation sections for locomotive stabling. Two separate transformers were required to enable trouble-free operation with the pair of hand-held controllers and common return wiring. This project evolved into an exercise in spaghetti and soldering. Controls for points, signals and trains were grouped together on one control panel that took many weeks of careful work and testing to complete. I now realize that DCC would have saved me a considerable amount of work; but it did not save the layout – it was scrapped when its shortcomings were laid bare by the adoption of DCC and the difference in approach to operation together with layout design.

As structural work on my second and more successful layout, Platform 4a & 4b, progressed, I was faced with a smaller version of the same wiring exercise and this led to a pause in the track and electrical planning process. How could I get the much needed operational flexibility from a layout with a scenic section measuring a mere 8ft in length without resorting to so much wiring? Good question!

At about the same time, I was toying with dc

lighting systems, having purchased a couple of kits from Express Models, particularly orange door lock lights for multiple unit stock. Whilst the systems work well, my research into DCC and decoders showed that there is a simpler and less costly route for locomotive and rolling stock lighting, one that would eliminate batteries and recharging from the equation, but with some different practical difficulties. All of these issues came together at about the same time and required a definitive solution before costs got out of control. Although battery- or capacitor-supported lighting systems offer constant lighting for dc (analogue) operation, the cost of installation became fairly prohibitive. I needed to increase the dynamics of the layout for exhibition use and to increase the play value for me, together with interest for the viewer. By all accounts DCC offered the most sensible solution to my layout control conundrum.

Only two wires (called the 'power bus') were required to get my exhibition layout up and running. Not literally, of course, because the layout was divided into sections to allow for wiring of handbuilt points with metal crossing vees and the need to feed power into the toe of such points to prevent short circuits – exactly the same as for analogue control. However, there are no cab control electrical blocks or isolation sections in the track layout and associated wiring running back to a switch panel via multi-pin connectors. The whole layout is live (think what that can do for coach lighting) and each track is supplied from two power bus wires that run the length of the layout from the back of the DCC base station. Each baseboard joint is bridged with a jumper lead containing just the two bus wires that run to tag strips. DIN plugs ensure that the jumper wires can be separated when the layout is dismantled for transport and storage. One tag strip under each baseboard provides a convenient terminal junction from which each individual running rail is supplied with current via dropper wires. It was up and running in a couple of hours and that layout operates today on the exhibition circuit, demonstrating the principles of DCC to anyone who wishes to have a go.

In the last six years DCC has completely changed my attitude to railway modelling and the way I look at layout design and concepts. When I buy a new locomotive, my thoughts turn more to making the lighting systems work prototypically, choosing the right decoder to obtain the best performance and wondering if there is room for a couple of speakers. When I design a layout, I no longer worry about the precise length of section blocks and where the control panel is going to be located.

In this book I offer a practical introduction to DCC for newcomers to the hobby as well as for experienced old hands. It is liberally sprinkled with a variety of modelling projects that can be used and adapted to suit different models. I have done all of these projects at my own workbench and made mistakes and discoveries on the way. I also feature a new layout project to demonstrate how straightforward it is to wire up a layout and incorporate some interesting equipment to enhance the operating experience.

Finally, I am keen to stress why I use DCC to operate my trains: it's all about achieving realistic operation, as close to the full-size railway as is possible, and about having fun whilst operating the layout. Gone are the cab control and electrical block switches. Gone are the fading running lights as a model comes to a stand. The complex wiring for analogue control is consigned to history as far as I am concerned. No more console control panels that look as if they could launch a mission to Mars. When driving trains, it's just me and the throttle, no different to the real thing.

**Nigel Burkin**
**September 2007**

CHAPTER 1

# What is Digital Command Control?

Digital command control is a great way of driving and controlling your model trains using computer technology that has been adapted for the task. In effect your layout becomes a networked computer system that uses small computers in your locomotives called decoders, which respond to instructions (data) transmitted from a central processor called a 'command station'. Each locomotive decoder responds to specific instructions sent by the command station using the decoder's unique address: this means you can drive trains independently of each other. This is a theme I will emphasize throughout this book – it's all about driving your trains. Nothing more than that.

After all, fun layout operation and being able to drive your trains in a realistic and flexible way is probably why you have taken an interest in DCC in the first place. It should be fun to use for all your friends and layout operators, too. When you speak to some DCC enthusiasts, however, you may be convinced that DCC is all about decoders, wiring, programming, complexity and digital data packets carried on strange alternating current waves through the track. Busses seem to come into it at times, even though train control is what we are after. Well, all of that is involved somewhere along the line – even a bus or two – but you do not need to know how DCC works to enjoy its benefits any more than you need to know how a jet engine functions in order to enjoy air travel. Some of the more complex stuff can be explored as your confidence in DCC grows, but all you will need to get started is in the box and described in the instruction manuals. DCC makes driving your model trains a fun thing to do and that's what this book is intended to help you achieve. All of the work in wiring up a layout, fitting decoders to your engines, adjusting the settings in a decoder and installing lights, sound and steam effects is a journey to the ultimate destination of realistic train operation.

## WHO INVENTED IT?

Wrapped up in the question of who could claim to be the inventor of DCC is the real one of where DCC originated in the first place and how it arrived in the world of model railways. DCC is not a model railway innovation but simply industrial machine tools technology applied to the problem of realistic model railway control. Märklin was the first company to

*Previously made awkward by the constraints of analogue control, double-heading (also called consisting) is made possible, with a refined level of control, by DCC.*

explore digital technology of this type (there were other attempts to do so) for model railway control, under the guidance of Bernd Lenz. After Lenz left Märklin, he further developed his ideas and established the company known as Lenz Elektronik GmbH. Most DCC users rightly regard Bernd Lenz as the 'father' of model railway Digital Command Control as we know it today. The application really took off, however, in the USA under the auspices of the National Model Railway Association (NMRA), driven by modellers' needs for a flexible control system that would allow multi-user operation on large layouts together with realistic multi-train operation completely unbounded by the constraints of controlling power in the rails through large panels of block switches.

Today, the far-sightedness of Bernd Lenz has ensured that certain of the system protocols essential for product inter-operability are published and available to all manufacturers, so the application of DCC is theoretically universal. Certain of those protocols are enshrined in the NMRA Standards and Recommended Practices (RPs). Not all areas are closely defined; there is much scope for individual manufacturers to demonstrate their technological flair and that leads to design differences in DCC systems that are not covered in the NMRA Standards and RPs.

## THE ROLE OF THE NMRA

The NMRA was established in 1935 as a gathering of model railway (model railroader) enthusiasts and other 'stakeholders' in the USA who were fed up with the wide-ranging standards from manufacturers for simple technical things taken for granted today, such as coupling standards, track scale and gauge. Prior to that, interchangeability of equipment between modellers' layouts was difficult and that was recognized as a bar to further development in the hobby. The NMRA devised sensible Standards and RPs to ensure that equipment from one manufacturer could be run with the equipment and on the track of another using the same power packs, couplings, wheel gauge and current requirements for a given scale. The basis of this system remains valid today: to be able to run N gauge equipment from Kato, Atlas, Athearn, Bachmann and Graham Farish together on the same 9mm Peco Streamline, Kato Unitrack or Atlas 'code 55' track is a given – not having such compatibility is now unthinkable.

The same thinking has been applied to DCC and the Standards are recognized internationally, not just in North America. The NMRA has devised standards and RPs covering such subjects as DCC electrical, DCC communication, DCC interface design and decoder wire colour,

*Running lights remain fully illuminated on a DCC layout, regardless of the locomotive speed. They stay lit even when a train is stationary and can be independently controlled at the push of a button.*

*Stabling locomotives together like this requires no special electrical blocks and associated switches. Drive a locomotive up to the next and park it.*

Configuration Variables (CVs) and DCC decoder transmission. In all, there are two Standards for DCC electrical and communications protocol to ensure that the DCC signal is the same no matter what command station you use, together with nine RPs. This means that any decoder from any manufacturer will work with any DCC system provided it is at least compatible.

Recommended Practices are regarded as only slightly less vital to operation than those considered as 'cast in stone' in the published Standards. Deviation is only permitted if there is a good technical reason for doing so. None the less, most manufacturers strive to comply with the RPs because the NMRA describes them as being 'only less mandatory than the Standards' and asserts that they are present to encourage maximum levels of interchangeability.

As far as DCC is concerned, the Standards and RPs only apply to the decoder and track power side of the system. This means that the DCC signals and associated areas such as decoder wire colours, certain CVs and digital communication are covered and must comply to the published NMRA documents to achieve a conformance warrant. It does not cover the controller (throttle bus) side of DCC systems, including electrical cabling that connects hand throttles, controllers, boosters or other devices to the system (this is often referred to as the system architecture). Nor does it specify design of systems in terms of buttons, connections, type of leads, plugs, sockets and so on. That is left to manufacturers' initiative and does mean that throttles and similar devices from one manufacturer may not necessarily work with command stations and boosters from another. The architecture for throttles differs between Lenz and Digitrax, for example, and you cannot make a Lenz throttle 'talk' to a Digitrax command station by simply plugging it in.

## NMRA Conformity Warrants and Product Compatibility

Conformance to the NMRA Standards and RPs is a good thing because it establishes a norm for all equipment, ensuring that if you want your friend's equipment to run on your layout, and vice versa, you can without compatibility issues. Those manufacturers that produce such products strive to maintain compatibility with the NMRA Standards and RPs because they recognize that a 'go-it-alone' approach would be commercial suicide. Some manufacturers, for various reasons, do not submit their equipment for NMRA conformance testing but still manufacture it to the Standards and RPs. Choose decoders and equipment that either hold an NMRA conformance warrant or are compatible and all should be fine, at least on the track side of the system that carries the digital signals. If the sharing of throttles is desired, then the architecture for the throttles must be the same (it is better to buy the same base system). But how do you know if equipment is DCC compatible or holds an NMRA conformance seal? And what is the difference between conformance and compatibility?

The NMRA not only writes and publishes Standards but has a programme to test products extensively for conformance to those same

*DCC compatibility symbol.*

*DCC conformance seal.*

Standards, RPs and any applicable industry norms. This provides a measurement of conformance to the published NMRA documents and those products that comply are allowed to display the NMRA Conformance Warrant Seal. The idea is that the NMRA programme provides modellers with some assurance of inter-product interchangeability both now and into the future. NMRA conformance should be highly valued when contemplating the purchase of DCC equipment together with products that are seen as compatible with NMRA Standards, such as those produced by Digitrax and ZTC. It is important to remember that the NMRA makes no guarantees, nor is it a DCC police force. It is well known that the NMRA provides guidance for manufacturers determined to be part of the DCC community with compatible or compliant products. The whole point is to ensure customer confidence in DCC and just one rogue manufacturer could really upset the apple cart.

Some manufacturers choose not to subject their equipment to testing but describe it as 'compatible'. It must be emphasized that products claimed to be compatible with the NMRA Standards and RPs have not undergone testing by the NMRA or any other organization that is currently needed for a manufacturer to claim conformance. What this means to you is that it is possible for a compatible product to work with other NMRA conformant equipment without difficulty, but some systems may not be totally compatible with other DCC products on the market. The symbol for compatibility is different to that applied to those regarded as being conformant to all applicable NMRA Standards and it refers to the ability of various products to work together. Whilst there are no guarantees, all DCC manufacturers stick to the spirit of the NMRA Standards. To deviate from them would not be helpful to any manufacturer and the DCC community quickly identifies

*Drive anywhere and park anywhere are important features of DCC operation. You soon become used to it and find analogue layout operation very inflexible after running a DCC layout.*

and avoids problem products that do not conform.

You do not need to know the exact details of the NMRA Standards and RPs to operate your chosen system. However, knowing the reason for their existence is extremely helpful in making purchasing decisions. Being aware that decoders from other manufacturers should work with your system is reassuring when contemplating an investment in DCC.

## WHY SHOULD I BUY INTO DCC?

Much work has been done to ensure product interchangeability, providing a measure of assurance to newcomers to DCC and a benchmark for established users to expand their systems in a coherent way. The question that many modellers ask is why should they adopt Digital Command Control technology in the first place when analogue has served them so well. What are the benefits? In other words, what's in it for me?

*Special lighting effects such as interior lighting and, in the case of modern multiple units, door warning lights can be installed and independently controlled using decoder functions.*

In the introduction above I described how my decision to adopt Digital Command Control for a new layout project completely revolutionized my attitude towards the hobby. It came at a time when my personal interest was waning and the thought of sticking yet more brass details onto another misshapen plastic model was too much to bear. Suddenly a completely new way of operating my layout and detailing my models was presented to me. I now routinely enjoy a number of benefits that are unavailable with analogue control. When exhibiting my layout at train shows, I never hesitate to demonstrate to others those very benefits that prompted me to make the change.

So, what are those benefits? Even before I explored the possibilities of digital sound, simple operational features became possible without having to throw a single switch on a layout control panel. One of the misconceptions about DCC is that it is only suitable for large layouts. On the contrary, the flexibility that it introduces to operations means that many of the operating constraints of small layouts are removed. Let us explore the benefits of DCC and what they offer to operating a layout of any size.

### Operational Detailing

DCC enables you to drive the train and not simply control the amount of power in the track. Put simply, an analogue controller only controls the amount of current in the track and the locomotive motor runs to the set power level. Nothing more. To obtain operational flexibility, it is necessary to fine tune the control of current in the track through banks of electrical block and isolating switches. It is possible to introduce operational sophistication, including shuttle modules and those designed to provide a degree of automated control for particular operating effects. But at the end of the day you are simply controlling the power in the track and all trains will respond to changes regardless of where they are and what device is regulating the current.

On the other hand, a DCC controller (throttle) is used to drive the specific locomotive allocated to it by keying in the locomotive decoder's unique address, which is decided upon by the modeller. It does not control power in the track, which remains at the same voltage no matter what buttons you press on the throttle or how fast your train is moving. By selecting the locomotive (or consist) address, you use the throttle to drive only that train, including its onboard systems, regardless of any other locomotive or train on the layout. Only your train will respond to commands from the throttle in your hand at that given time. You can control the running lights according to operating (driving) conditions. Digital sound decoders mean you enjoy the real sound of your favourite engines, sounding the horn or blowing the whistle at whistle-boards and sounding the bell at grade crossings.

*DCC is all about layout operation and enjoying it, too! My fellow modeller and layout operator, Eddie Reffin, demonstrates how enjoyable operating with a hand-held throttle can be to an exhibition visitor and they both appear to be having fun, judging by the smiles!*

This is a layout operating experience more akin to driving a real train: one operator (driver or engineer) per train with one set of controls (which is why the hand controller is often referred to as a 'cab' or 'throttle') unique to that train until control is given up, say at the end of a run. You can operate from one position or follow it around the layout, doing shunting or making station calls; obeying signals and following the rules of the road to avoid collisions with other trains being independently operated by fellow operators. All without section switches!

In the USA, large layouts are operated by following a train around the layout on specific assignments, sometimes operated to timetable, train order and fast clock practice or as 'extras' or using Centralized Traffic Control (CTC) rules. There may be an individual assigned to signalling or train dispatching duties that include signal control or CTC if applicable – depending on the country of operation. There may also be operators assigned to yard switching or shunting and to hostling or depot management. It all very much depends on the size of your layout and the number of operators at your disposal.

## Constant Lighting Effects

One of the frustrations of analogue control is the difficulty in controlling running lights in locomotives and rolling stock. When a locomotive equipped with the usual factory-installed lighting is brought to a stand, the lights go out. This is because the power in the track is switched off to prevent further movement, which means there is no power available for lighting. There are some very clever ways of ensuring that lights can remain illuminated for a period of time after the power has been cut, even without the use of onboard power sources such

*Operation is all that DCC is about: recreating scenes like this in model form. Four large locomotives head a long coal drag up Bozemann Pass, Montana, in June 2007. This is not impossible to recreate with analogue control, but very simple with universal or advanced DCC consisting!*

as batteries. However, these are limited in their application and the circuitry can take up a great deal of space. The expense of such electrical circuits may also be similar to the purchase price of an advanced DCC decoder!

The rails of a DCC layout are constantly live and this means that the onboard decoders are in constant contact with power and the command station. When a train is brought to a stand, it is not done by changing the voltage of the track power. Lighting systems are constantly supplied with current, which means that lights remain illuminated regardless of train speed until they are switched off by pressing a button on the throttle.

Many decoders are equipped with components and software that enable them to provide a variety of lighting effects including strobes, flashing, constant illumination, dimming and the ability to adjust brightness of lights to suit operational requirements. The running lights of a locomotive can be switched on and off at will and individual lights can be controlled using different function buttons on the hand throttle. Suddenly, those flickering, dim and unrealistic lighting effects come alive with DCC, making the best use of factory-installed lighting in modern models.

As the modeller becomes familiar with the wiring of decoders, the opportunity to explore different lighting effects becomes irresistible. Why not equip your model with cab lights, independently controlled tail lights, lights that comply with US Rule 17 on US-outline models or switch over from day to night-time lighting positions on modern British diesels? Oil lamps on steam locomotives can also be represented together with onboard passenger coach lighting, flashing tail lights or constantly illuminated oil lamps on the back of trains and built-in tail lights on vehicles such as mail vans and sleeper coaches. Excited by the possibilities?

## Sound Effects

Digital sound in model locomotives is fast becoming the biggest selling point for DCC, even though dc sound systems are available, particularly in US-outline models. Digital sound is recorded on special decoders that control the delivery of sound effects (through an amplifier) in three ways: driver-controlled at the press of a button on the hand throttle (horns, bells, whistles); controlled by the speed and direction of the model as determined by the chosen speed step (engine note, brake squeal and the like); or random sounds can be provided that are not controlled by the operator (compressors).

A sound-equipped diesel locomotive model can be 'started up', driven off, brought to a stand and shutdown with sounds from equipment such as horns, bells, air brakes and compressors. Suddenly you are driving the model just like the real thing! Once experienced, never forgotten.

*Digital sound effects can be built into ready-to-run locomotives such as the Bachmann Class 37/4. Increasingly, manufacturers in the UK and the USA offer locomotives with digital sound as off-the-shelf products. 37 406 was making plenty of noise as it tackled the sharp inclines of the Rhymney line with a service from Cardiff Central on 26 April 2004. Fancy modelling that in 4mm scale, complete with the sound effects? You can do it with DCC.*

*EWS Class 66 No. 66 064 passes Pewsey with the weekly Exeter Riverside–Dollands Moor freight on 28 June 2004. The sights and sounds of a heavy freight train can be created in model form using digital technology, including prototype running lights and the distinctive 'ying' of an EMD diesel engine. Bachmann has also offered this type of locomotive, equipped with lights and digital sound installed at the factory, for as little as £150–160 off the shelf.*

## Drive Anywhere, Park Anywhere

It dawned on me very early in my dabbling with DCC that it provides a great deal more operational flexibility than analogue cab control. You may wonder what I mean by this. When you operate a DCC layout for the first time, you will be surprised at how simple it is to drive your chosen train, be it a single locomotive or a heavy freight train to any point on a layout. Suddenly there is no need to locate and throw sections' switches on complex control panels or endure the embarrassment of seeing your train come to an abrupt halt as the result of forgetting to switch a section to your controller. This is called 'drive anywhere, park anywhere' operation, because isolating your train from others during an operating session to prevent unwanted movement no longer applies. This becomes particularly noticeable when stabling locomotives at motive power depots and in sidings. In that situation, there are no isolating sections in the track to remember. Drive your locomotive up to the next, stopping it buffer to buffer if you really want and forget about locating the right

*As the EMD Class 66 locomotive has been bought in numbers by UK operators, changes in the running light fittings have been made to reflect modern operating practice. This includes the use of LEDs for marker lights that have a blue tint to them, together with combined forward and reverse marker lights in one housing. All of these lighting effects can be modelled simply and accurately with DCC using a decent decoder, a handful of quality LEDs and some resistors. An intermodal service hauled by GBRf Class 66 No. 66 710 passes the site of Williton power station shortly after leaving the Sheet Stores Junction line from Trent on 9 July 2004.*

isolating switch because there isn't one. What a relief that is!

## Double Heading (or Consisting)

Remember how much hassle it is trying to get two engines to run together perfectly on dc control? Even arranging the formation of a double-headed train at the depot is not always as easy as it is on the full-size railways. DCC offers two consisting features by which you can double-, triple- or quadruple-head your trains with ease and operate them together with just one address. US-outline modellers find this feature particularly beneficial.

## Simplified Layout Wiring

The wiring for a DCC layout can be as simple or as complex as you need it to be, but it is still infinitely easier than analogue cab control. Whilst the idea of connecting just two wires to one-piece track to operate a complete layout is not strictly true, the power supplied to the track is carried by the power bus, which consists of two beefy cables that run the length of the layout. Power is supplied to each piece of rail through dropper wires connected to the power bus cables. Together with the controller or throttle bus (assuming you wish to have different controller points for multiple operators), a two-wire power bus is the minimal layout wiring that you

*DCC offers operational flexibility to a layout. Locations such as stabling points, depots and stations where space is at a premium benefit from the ability to drive and park locomotives anywhere – buffer to buffer if necessary. Once you have parked your engines, you could leave the lights on, too, if you wish! This is King's Cross stabling area with two locomotives in attendance when photographed on 2 April 2005.*

need for a fully operational DCC layout. Turnout control requires additional cables if they are to be electrically operated from a separate panel; that wiring will be independent of the power bus cables and will need a separate power supply. Turnout operation (and signals) can be simplified if they are controlled through accessory decoders, which will draw their current from the power bus or separate power supply. If sheer simplicity is your thing, consider hand-operated turnout control (and many modellers in the USA choose to use ground throws to switch turnouts, especially in yards and depots) to keep layout wiring simple.

With my personal preference for small, portable exhibition layouts, not having to wire in a multitude of section switches to obtain operational flexibility with more than one 'train in steam' was a huge relief and saved a considerable amount of money. The wiring in my layouts is particularly simple as a result, making construction quicker and easier. After all, I want to operate my layout as soon as possible, not spend any more time than I have to at the workbench with soldering iron and reels of cable!

## SHAGGY DOG STORIES: WHY YOU SHOULD LAUGH AND THEN MOVE ON!

Sadly there are a number of urban myths surrounding DCC. It's one of those subjects that cause much debate between the supporters of analogue control and those who prefer the digital route. Flame wars on Internet forums only demonstrate how polarized the debate can be, eclipsing the P4, EM and OO gauge controversies that once dominated the hobby in the UK. Why some modellers wish to discredit DCC as a control system may be down to a number of factors, of which cost and an unfounded fear of technology may be two. The result is the spreading of misinformation,

*BNSF 6817 and BNSF 6819 were allocated to helper service out of Essex, Montana, when photographed in July 2007 pushing at the rear of a long train of autorack freight cars. Operational realism was the driving force behind the adoption of DCC as a model railway technology. Imagine trying to control helper engines on banking duty with analogue control. Working the electrical block switching to maintain independent control of two sets of locomotives, one on the front and one pushing, would be quite challenging. By using DCC control, two operators can recreate helper operations like the one over Marias Pass on the BNSF Hi-Line through north-west Montana. DCC enables the independent operation of the lead engines and the helper set in the same way as the full-size railway does and without having to throw a single electrical block switch.*

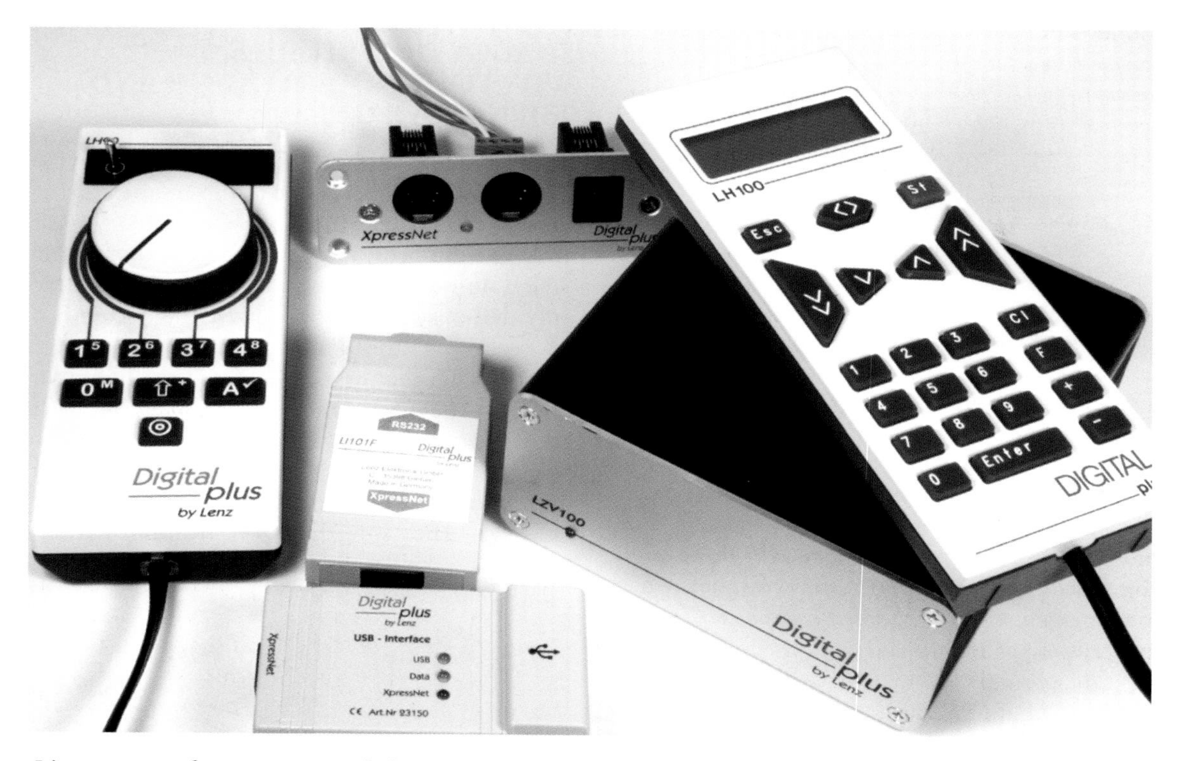


*It's so tempting for newcomers to dash out to the shops and buy one of everything from the catalogue. Avoid buying equipment that you are unlikely to need to get started, because getting to know your system from scratch will be enjoyable enough and extras can be bolted on later as confidence and knowledge grows.*

including stories that are now regarded as urban myths of some importance within the DCC community. In fact, there are so many that they could fill a book. Here is a selection I have heard from nay-sayers. Should you be unfortunate to bump into one quoting any of the following popular 'urban myths', laugh, then move on and enjoy your digital system.

**1. Digital decoders reset their address to default when a locomotive is run round a continuous loop layout**

This myth was perpetuated by a well-known exhibition trader in the UK for quite some time. I once overheard it repeated by people at a train show while, at the same time, they watched a continuous-run layout operating with DCC control. Fortunately no one has told the decoders.

**2. Car light bulbs are needed to provide short-circuit protection**

In the early days, car light bulbs were used as insurance against damaging short circuits and the idea made a great deal of sense at the time. However, that is no reason for not buying into DCC. Modern systems have such sensitive short-circuit protection that the old light bulb trick has vanished into the realm of legend.

**3. DCC only benefits large layouts**

A commonly heard comment is 'My layout is only small, so DCC is no good for me'. My experience with compact portable layouts demonstrates that DCC control dramatically benefits small layout operation and can actually enhance the driving experience beyond recognition. If you are planning a compact layout, you could use an entry-level DCC set to save money rather than invest in one of the advanced starter sets such as the Lenz Set 100. However, keep in mind how your plans may evolve over time – you may need more capacity and functionality in the future!

### 4. I must know how to programme computers to make DCC work properly

It is unfortunate that the word 'programming' has been applied to the simple task of changing settings in a decoder (changing the CVs). I no more know the principles of computer programming than the next person. Such knowledge is completely unnecessary to enjoy all of the benefits of DCC. In fact, some people who work with computer programming are surprised to find that their programming skills are completely unnecessary. These days, with simple menus designed to simplify 'programming' built into systems, hexadecimal (HEX) calculations are increasingly unnecessary unless really advanced programming is required. HEX calculations cause the typical modeller (including me) to run screaming for cover at their merest suggestion and so it is a relief to use those systems with conversion tables and the ability to enter easy-to-understand decimal values for HEX values.

### 5. The myth of the DCC-friendly point or turnout

I do not know where this one came from but there is no such thing as a DCC-friendly (or unfriendly) turnout. If turnouts work with analogue control, they will work with DCC. After all, it's only electric current that is flowing through each rail of your track, be it dc current with analogue control or the alternating current (ac) signal for DCC. The same principles regarding short circuits and other properties of electricity still apply. If your turnouts have metal crossing vees, they must be isolated from power feed from the diverging end of the turnout, using insulating rail joiners. If the turnouts you are using are not power-routing types the power supplied to a metal crossing vee must be polarity switched when the turnout is changed, to avoid a short circuit regardless of the system you are using. Some manufacturers' turnouts benefit from simple modifications to enhance their performance with DCC and that will be discussed when we tackle layout wiring.

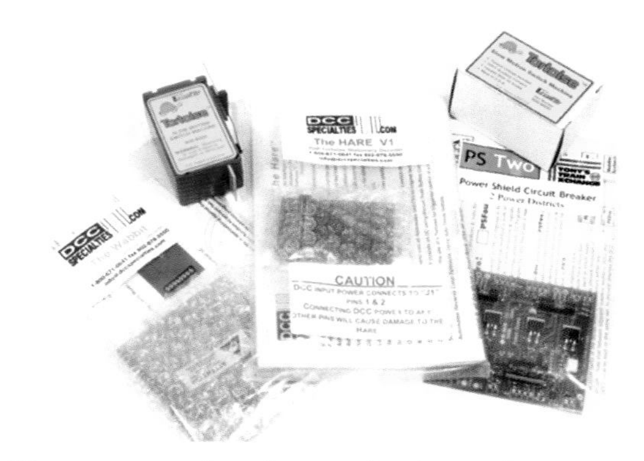


*There are a number of companies that produce accessory equipment for DCC systems, such as circuit breakers, specialized and programmable accessory decoders and similar devices.*

### 6. You will need a decoder testing device

No, you do not need to spend valuable modelling funds on complex decoder testers, nor do you need a personal computer to programme decoders if you do not wish to. It's an interesting branch to DCC and one you may wish to explore in the future, but those modellers who prefer to drive their trains rather than spend modelling time sat in front of a PC will be delighted to know that it is not essential. All advanced digital systems will enable you to test your decoder and its installation in a locomotive through the use of a 'service track' and programming features built into the system. That said, a simple decoder tester that works as a 'stationary' locomotive for programming and testing lighting effects is a useful device to have to hand on the workbench and building one for yourself is an interesting project that is described in the next chapter.

**7. You have to buy expensive boosters or it won't work**

Boosters are devices designed to provide additional power when the layout is divided up into power districts (boosters may also be called power stations and amplifiers). To begin with, all DCC systems have a booster included in the base station or package (somewhere) to get you started. If you run your layout solely with the supplied base station, it automatically becomes a single power district layout. Adding further power districts offers many benefits rather than being a disadvantage. By adding a second booster to create a second power district to separate a layout feature such as a large freight yard from the mainline means that there is more power available for trains to operate and greater short-circuit protection. A derailment in the freight yard causing a short circuit will shut down that particular district, leaving the rest of the layout operational. A separate booster can be used to supply power and DCC signals to accessory decoders via an independent power bus circuit for turnout control ensuring power for the trains is not reduced by the needs of setting turnouts. This arrangement protects turnout and signal control from short circuits arising from derailments and turnout run-throughs and is actually very useful indeed.

**8. You have to keep your track very, very clean or DCC won't work properly**

Now that one came as a surprise to me when I was told that as a reason not to buy into DCC.

*The digital train set offers an entry point for younger modellers and those experienced modellers who wish to try DCC without disrupting a stable operating environment on an existing analogue layout. In the highly unlikely event that the modeller does not enjoy operating with DCC, the financial outlay will not be as high as buying an advanced starter set. However, small entry-level systems like this one are limited in their scope and the modeller is likely to grow out of it very quickly indeed!*

*The Lenz Compact is a typical entry-level DCC system with an impressive set of features. Like most budget and entry-level sets, it does have limitations when compared to advanced systems such as the Lenz Set 100. Limitations can include a reduced power output (typically 2.5 amps), a limit on available locomotive addresses (usually 0–99 or 0–127) and a layered menu system to access functions such as accessory decoder control.*

*Some entry-level systems have an impressive specification including four-figure address capability and other features normally found in high-end systems. However, when the manufacturer is pegging the price to less than £100, including the amplifier/booster together with a suitable transformer, something has to give! In the case of the Roco MultiMaus, the system has some great capability, including compatibility with the Lenz system, but has limited power and no independent service track output. When used with a Lenz command station, it performs like any Lenz hand throttle with CV readout and programming features.*

The guy who repeated it to me was watching one of my DCC layouts in action at a train show on a Sunday afternoon, although I do not think he was aware that the control system being used at the time was a digital one. The layout track had been cleaned during the Friday set-up and not touched since then, with no operating problems. Of course, this myth is nonsense. Track has to be kept reasonably clean for reliable operation, regardless of analogue or digital control, but certainly DCC layouts require no more attention than any other.

## TRY BEFORE YOU BUY

One of the best things that newcomers to Digital Command Control can do is to gain some first-hand experience on a layout for themselves before deciding to buy. Do not pass up the opportunity to visit a layout already equipped with digital controllers to see if it really suits your operating style and test its features to see the benefits for yourself. It is very difficult to visualize the fundamental change that DCC will bring to the operation of your model railway until you hold a throttle in your hand and in direct control of a train. One thing that will become immediately apparent is how easy it is to operate a digital layout and how enjoyable. It will bring it home that the whole principle of DCC is to drive trains as realistically as possible. The second thing that will become apparent is that all the locomotive lights remain illuminated at all times, even when locomotives are at rest. If the owner of the layout you're visiting has locomotives equipped with digital sound, take the opportunity to drive one and explore this third dimension in model railway control. Sound is very evocative and to operate models of your favourite locomotives accompanied by authentic sound is quite special, especially when you experience it for the first time.

At this stage it is irrelevant which system is being used on the layout because you are there just to test the principles of DCC and to satisfy yourself that it will meet your needs and deliver the benefits that you seek. Product research can follow your first taste of driving with digital control.

## WHAT SHOULD I BUY?

There are numerous systems on the market, some from the USA, some from Europe, and one British one, too. Making a sensible choice, like any consumer decision, is not easy. However, given the investment required in terms of capital outlay and time, it is worth spending some time researching each manufacturer and its products. As stated earlier, NMRA compliance and conformance should be an important consideration, as should the ability of the system to be upgraded and improved. This might mean purchasing an advanced system straightaway even if your collection of locomotives and layout are small. Inevitably, model collections grow and the opportunity to build a larger layout may present itself in the future. Whatever happens, try to avoid wasting money and effort by selecting those systems that are structured for growth and offer the means of protecting your original investment.

The following points are worth bearing in mind when choosing a system:

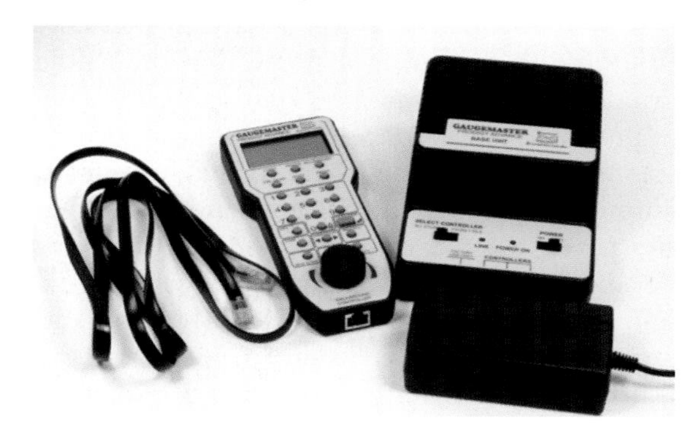


*One of the easiest intermediate to high-end systems to use is the Gaugemaster/MRC Prodigy Advance, which comes as a package including a base station, hand-held throttle and transformer. Its simplicity compared to some advanced systems has seen it make inroads into the UK market.*

- The locomotive fleet: its size and the need for special effects such as digital sound and lighting will determine the choice of system and whether its hand throttles will support the required number of functions.
- Size of layout: it is important that the system can be expanded with additional boosters and other equipment as your layout grows. Small layouts also benefit from advanced systems because features such as a computer interface and two-way communication are just as applicable.
- Desired operation: do you plan to run from one control point or intend to follow your trains around the layout? This may determine the type of throttles you need and whether the system should offer connection units (extension plates) that can be built into the front fascia of the layout. Wireless or infrared control may be important if roaming operation without a tether is your goal. Being able to plug in more than one hand throttle will make operating sessions with your friends that much more fun, too, so check the method that various systems use to offer multiple operator control, such as throttle bus extension plates.
- Number of operators: this will determine how the throttle bus can be expanded to accommodate comfortably more than one operator around a layout. I will discuss the use of extension plates to expand the throttle bus later when the method of wiring a layout for DCC is demonstrated.

## ENTRY-LEVEL SYSTEMS

Entry-level systems are offered as an economic way to take the first step on the DCC ladder. Such systems are commercially priced to be attractive to modellers with small collections and small layouts or those unwilling to make a large investment at the start. Like all things in life, you only get what you pay for and basic starter systems rarely have the full functionality of advanced starter systems. They are ideal for shunting plank-type layouts and small compact exhibition layouts that rarely need more than a handful of locomotives.

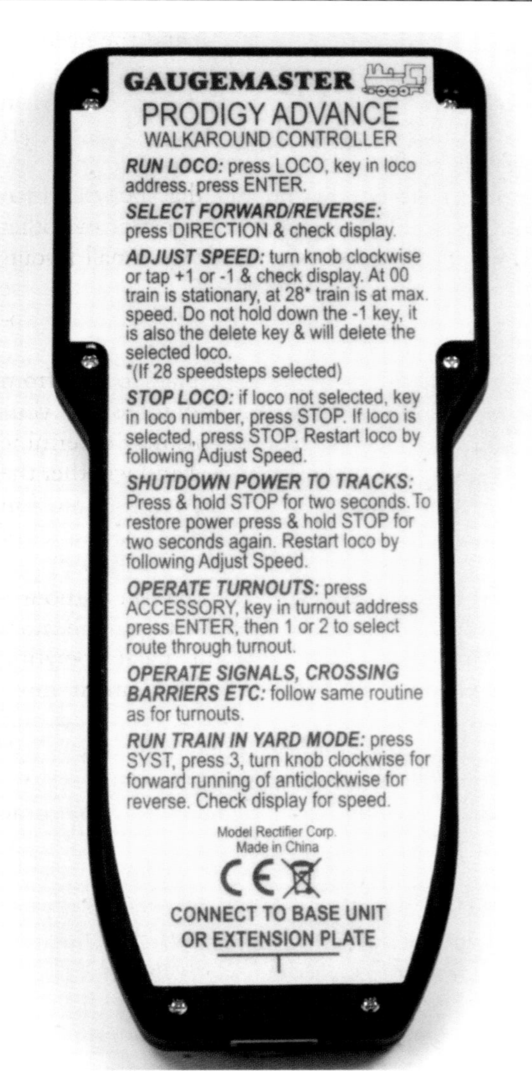


*Helpful instructions for the main functions are printed on the rear of some system hand-held throttles, which is really helpful when in the middle of an operating session. This is the rear face of the Gaugemaster throttle.*

Whilst there is a great deal of difference between entry-level systems, they typically offer 2-figure addresses for your locomotives, from 0 to 99, although some very basic systems may offer as few as ten locomotive addresses. They may not offer a programming track facility, so programming has to be done through the layout using a 'broadcast' method, which throws up all sorts of practical issues as your involvement with DCC becomes more advanced, nor may they have more than four function buttons. They may be limited in their overall functionality and power output, which is typically about 2.5 amps, although some are restricted to as little as 1 amp. They combine the command station, booster and throttle into one package, which means they can be awkward to use on some types of layout.

None the less, do not disregard such systems because some do offer a throttle bus port for the connection of hand throttles, such as the Lenz Compact. Others are very smart in combining the command station with the hand throttle, so walk-round control is possible (NCE Power Cab). Others offer 4-figure addressing and other advanced features. If it is unlikely that you will ever need advanced digital features such as computer control, RailCom two-way communication, 4-figure locomotive addresses or more than four functions, a top-end entry-level or intermediate level system may suit both your environment and your pocket.

Many manufacturers are aware that there may be need to expand the system at a later date and most entry-level sets are designed to be

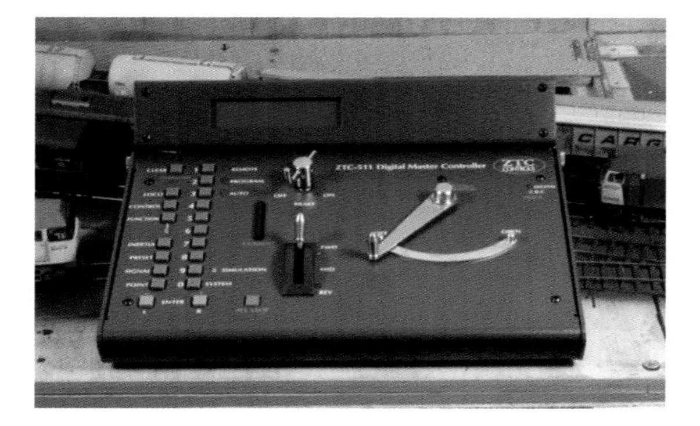

*A typical command station and booster combined with a set of driving controls is demonstrated by the powerful ZTC-511 master controller. This type of DCC base unit is called a 'console'. It can be added to by attaching hand-held throttles that ZTC calls slave controllers.*

incorporated into advanced systems from the same manufacturer, thus protecting your original investment. For example, the NCE Power Cab becomes a Deluxe Pro Cab when connected to the advanced NCE system. The Digitrax Zephyr and ZTC-505 are top-end entry-level systems that are ideal companions for the advanced sets, which is helpful as the modeller expands their system. Many of the entry-level base stations and hand throttles turn into companion controllers when plugged into advanced DCC base stations. The actual device may need to be switched from main controller status to 'slave' status: the manuals will advise on how that is done. Before you make a purchase, read the design specifications carefully to ensure that you're not buying into a dead-end system that will have to be later discarded when expansion is required.

## ADVANCED STARTER SYSTEMS

Advanced starter systems usually include the command station, booster and one hand throttle together with associated cables and manuals. Some may provide a suitable transformer, such as that supplied with the MRC/Gaugemaster Prodigy Advance. It is not uncommon to find both the command station and booster incorporated in the same box, called a base station. They usually have full DCC functionality such as four-figure addressing, multiple function control, computer interfaces, two-way communication, transponders, signal and turnout control together with a higher track power output, typically 5 amps for HO/OO gauge. Product research is vital to ensure that you purchase a system with the features you need, because they vary widely between different manufacturers' systems.

Advanced systems are ideal for large locomotive fleets where address options beyond 0–99 or 1–127 are required. Those layouts (of any size) that will eventually be equipped with digital signal and turnout control, a computer interface, occupancy detectors and feedback modules will benefit from the increased functionality of an advanced system.

Advanced starter systems are designed to get you up and running as quickly as possible with something that can be expanded as confidence in and knowledge of the system grows. At the very beginning the only items you should contemplate purchasing are an advanced starter set, a suitable transformer (if it is not already supplied as part of the package) and a handful of locomotive (mobile) decoders.

## BEFORE YOU TAKE THE PLUNGE

It is tempting to acquire a catalogue of your preferred system and to make a shopping list of everything you think you will need, such as a computer interface, two-way communication devices, Uninterruptible Signal Processing (USP) units such as the Lenz Power-1, turnout and signal control decoders (called accessory decoders) and block management modules. It's easy to run up a shopping list that would make a massive hole in your bank balance. Resist the temptation to do this because, if you are unfamiliar with DCC, just learning the capabilities of the base station and throttle together with locomotive decoders is going to be a sharp learning curve.

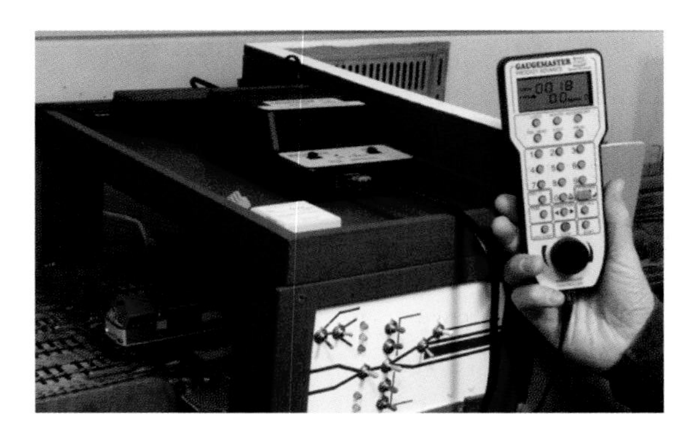

*This exhibition layout normally operates with a Lenz Set 100. However, because manufacturers are seeking NMRA conformance warrants, already have the warrant or are developing their equipment to be compatible, most, if not all, DCC equipment can be connected to this layout's power bus wires and locomotives equipped with a variety of decoders safely operated on it.*

There's plenty of time to explore the rest later, as I have discovered in the last few years.

## GO SHOPPING!

At the beginning your shopping list should be relatively short. Either buy an entry-level system that comes complete with a dedicated power supply together with some decoders or, if you think you may grow out of that very quickly, take a look at advanced systems.

When contemplating an advanced system, you will need a starter set that consists of a base station (booster or amplifier and command station in one box) together with a hand throttle capable of undertaking all programming functions. You will definitely need to buy a separate transformer because they are rarely supplied with an advanced starter set: go for a 5 amp power supply, which will provide sufficient power for N gauge, OO gauge and HO scale. Large scale modellers may need more power than 5 amps, and 8 amp power sources are usually preferred. Check that your particular system will support 8 amps or consider boosters and power districts instead.

If you are unsure of how much power you will require, do a simple calculation by adding up the likely power consumption for the total number of locomotives present on the layout during a typical operating session. Whilst it is unlikely to exceed 5 amps for OO/HO gauge, very large layouts may need more power than that and the extra can be introduced by adding a separate booster.

Allocate a power consumption value to the maximum number to be operated at any one time (around 0.5 to 0.75 amps per running locomotive in HO scale; 0.3 amps for N gauge locomotives). Also factor in a value for decoders in stationary locomotives, which are taking a tiny amount of power from the system at all times, even when the running lights are extinguished, and add that to your power consumption figure. It's not unreasonable to add about 0.1 amps for each stationary locomotive with running lights illuminated. Finally, do not forget the power consumption by lighting systems in coaching stock and also any power that may be used by accessory decoders.

Finally, your shopping list should include a handful of mobile decoders.

*Decoders are essential and should be on your shopping list. However, with so many locomotives being offered with DCC interface sockets (called 'DCC-Ready'), they are increasingly becoming plug and play devices to the extent where some decoders, such as this Lenz Silver Direct, don't even have wires.*

CHAPTER 2

# Opening Pandora's Box

It's exciting! You have just engaged in a satisfying bit of retail therapy and you are probably dying with impatience to dig all those interesting control boxes and cables out of the packaging. The very first thing you should do is read the manual, at least the first part of it that describes the system and how it is connected together. A good way to become familiar with your system before connecting it to a layout is to place a couple of yards of set track on a table and connect the system to it. Equip a couple of reliable locomotives with decoders, ideally those fitted with a NEM DCC interface socket and lights, and play with them for a couple of hours. Run them together, singly and coupled as a pair to become familiar with the function buttons, speed steps, emergency stop button, multiple working (consisting) and how performance differs over analogue control. Have the manual for your throttle to hand so you can refer to it.

There is something fascinating about running two models in close proximity without having to use an isolating switch to electrically separate them. When I bought my first DCC system I did this very thing, learning about my new system with a few lengths of track and a couple of locomotives. I hadn't had so much fun with model trains in years than I enjoyed that evening

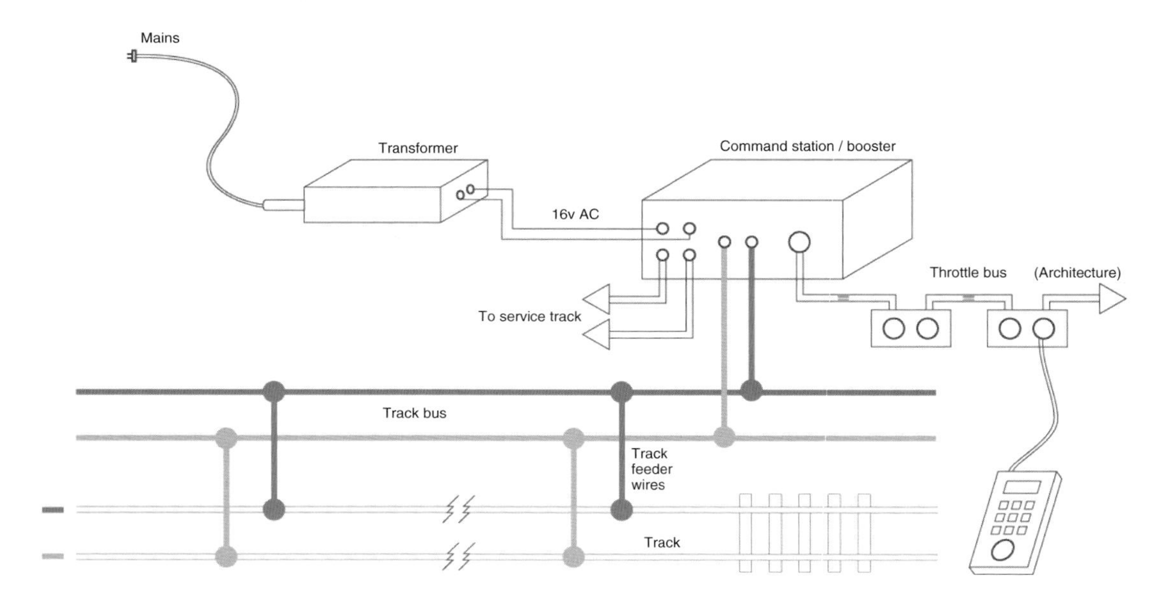


*Typical DCC structure showing how the various components go together.*

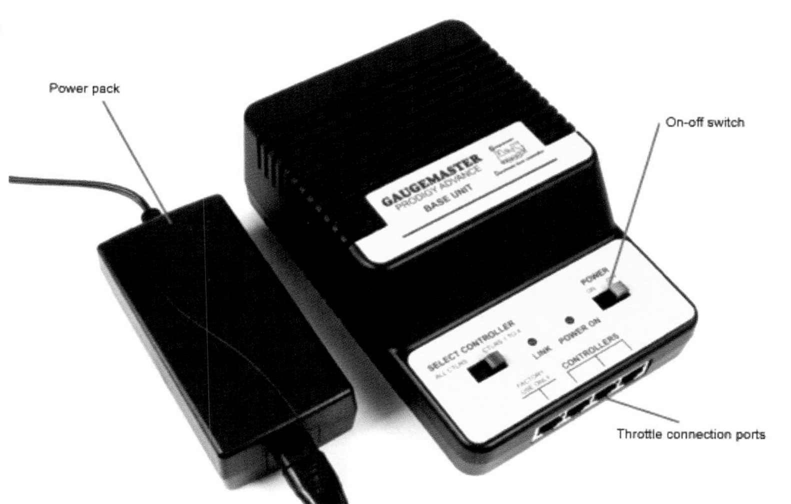


*A typical command station and booster combined in the same equipment casing (called a base station) produced for the UK market by the Model Rectifier Corporation (MRC) and marketed by Gaugemaster.*

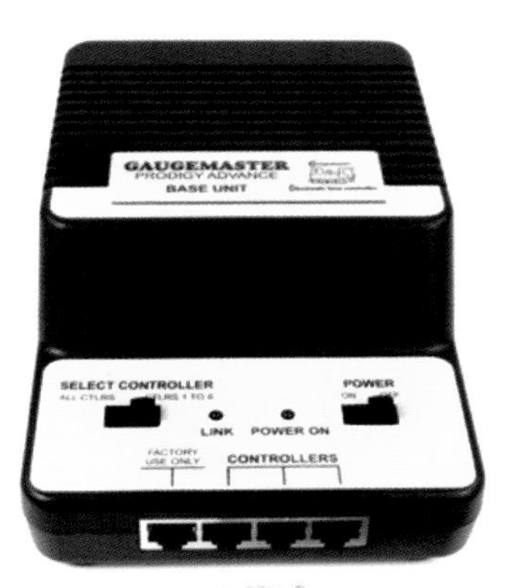


*A front view of the Gaugemaster/MRC base station. There are three ports for connecting throttles; most systems only offer one port for a throttle connection in the command station/booster casing.*

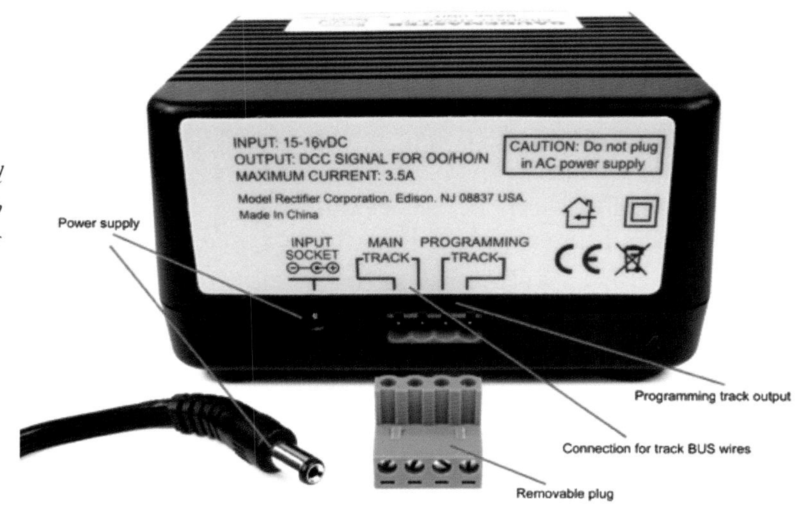


*The ports at the rear of the Gaugemaster/MRC DCC system are fairly typical. The green plug makes it easy to disconnect the base station from a layout should it be needed elsewhere. Such plugs are usually supplied as spare parts and each set of layout bus wiring leads can be fitted with one.*

when I took my first tentative steps into DCC. The rest, they say, is history.

Before you start, some understanding of the structure of what you have bought is useful, but not essential, to setting up a successful DCC layout. However, you should be aware of certain facts about track power and how signals are sent to the locomotive decoders. You will be pleased to know that high order programming will not be on the menu; you simply do not need to know anything about computer programming to enjoy the benefits of DCC.

## SYSTEM COMPONENTS AND HOW THEY WORK

A typical DCC system is composed of four key elements that play a specific role in making the system work. I have alluded to them in Chapter 1 without explaining what they do. Now you have the box open with those sophisticated-looking devices littered across the table, you may be wondering about the role they will play in the operation of your trains. Here's a brief explanation.

### Command Station

The command station is the intelligence centre or brains of the operation, the glue that holds it all together. It is a computer that contains a micro-controller and stable memory enabling it to control communication between the controller (throttle) and locomotive decoders. Communication between the command station and decoders takes the form of data packets, which are made up of a stream of binary digits. Typically, about 200 packets per second are transmitted to all decoders in contact with the command station, keeping them fed with data packets many times a second.

As far as each decoder is concerned, it receives many specific data packets bearing its address with speed and direction instructions every second, which are all the same if there's no change on the throttle. The high frequency of packet transmission is important to how the system works. For example, should a data packet containing a speed change instruction for a specific locomotive be interrupted by dirt or 'noise', it will only be a few micro-seconds before it is sent again. The delay in the locomotive responding to the instruction change is tiny and cannot be detected by a human operator.

### Booster

A booster (also called a power station or amplifier) literally boosts the power together with the digital packets from the command station so the track voltage is high enough to run model trains. The output signal is an alternating current (ac), but nothing like the sinuous waveform associated with domestic electricity supplies. The DCC signal is a square wave set to about 14V ac to 16V ac at a frequency of about 8kHz.

The signal is powerful enough to overcome the disadvantages of transmission through metal rails with all of the electrical 'noise' and physical dirt common to model railway layouts. The decoder can extract the data packets as well as rectify the current to provide power for the motor and onboard systems such as running lights. Power is usually supplied to the motor as a Pulse-Width Modulated (PWM) dc current. Think of the power station or booster as the 'beef' behind the brains of the command station; it contains no intelligence. It's normal for both the command station and booster to be combined in the same casing, sometimes called a 'base station'. Additional boosters can be purchased to create 'power districts' and to boost power supplies when large collections of locomotives are present on the layout.

### Throttle

The 'throttle', which is also called a 'cab', is the train 'controller' equipped with a keypad and/or control knob, an LCD or illuminated display screen, decoder function controls (function buttons) and decoder programming controls. Throttles are usually hand-held, although entry-level sets may incorporate the command station, booster and throttle into a desk-style unit (Lenz

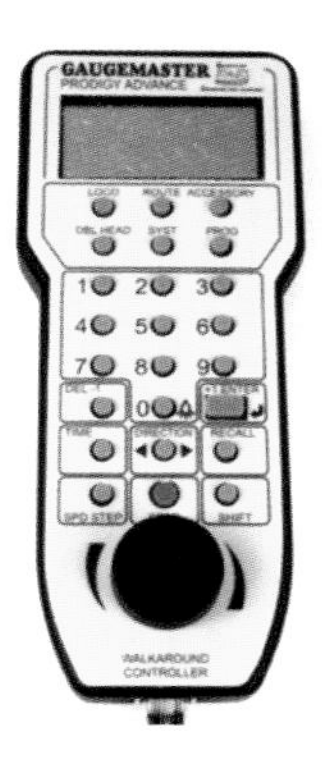

*The throttle (or cab) is the control desk of your locomotive. In other words, the driving controls. It also provides the keypad for programming decoder CVs together with a readout of CV values on the LCD screen.*

Compact, ZTC-505 and Digitrax Zephyr) or hand-held unit (Roco MultiMaus and NCE PowerCab). Some very advanced systems also offer a master controller console with controls, so there are no hard and fast rules regarding the exact format.

The locomotive control element is called a throttle or cab because it represents your locomotive driving control desk or stand. It has all of the controls necessary to drive your train and control onboard systems. Typically, you need one throttle per layout operator and they are usually designed for roaming, enabling an operator to follow a train around the layout, disconnecting and reconnecting the throttle at various points around the layout. Interestingly, a locomotive will continue to obey the last instructions from a throttle, even when the throttle is unplugged from the layout or a different locomotive is chosen. The fact that the throttle does not control the track power is illustrated when, in cases where the system base station is not incorporated in the throttle, a model continues to run even when the throttle is disconnected from the layout.

Wireless and infrared technology has made the roaming experience more enjoyable and without the panic of searching for a socket (on unfamiliar layouts) to plug the throttle back into the system to control a moving train when changing locations on a large layout.

There is more on the driving experience in Chapter 7, including the setting of values in the decoder (Configuration Variable programming) to refine the operating characteristics.

## Decoders

Decoders are awesome in their power when compared to their physical size. They extract the digital packets from the digital signal, processing the data and obeying instructions contained within the packets. They also rectify the ac

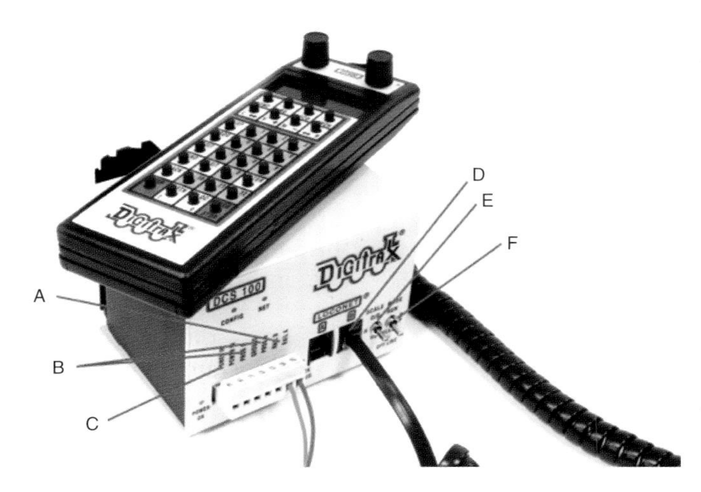


*An advanced DCC system showing the base station and associated throttle. Note that base station ports and controls can vary between systems. For example, the connections for the power bus (A), service track (B) and power supply (C) are located in the front of the base station in the case of this Digitrax unit. It is a convenient unit to use because the throttle bus ports are also located in the front (D) together with gauge voltage controls (E) and a power switch (F). The Digitrax Super Chief is one of the most advanced DCC systems available.*

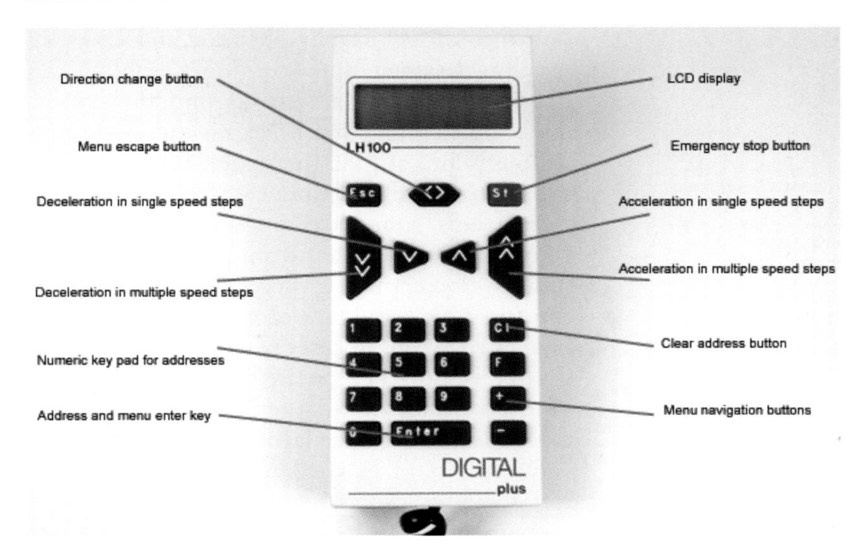


*The Lenz LH100 throttle is a good example of a hand-held throttle that uses buttons for speed control. Some operators prefer buttons, others prefer control knobs. You soon become accustomed to using throttles and associating buttons with driving controls in a locomotive cab.*

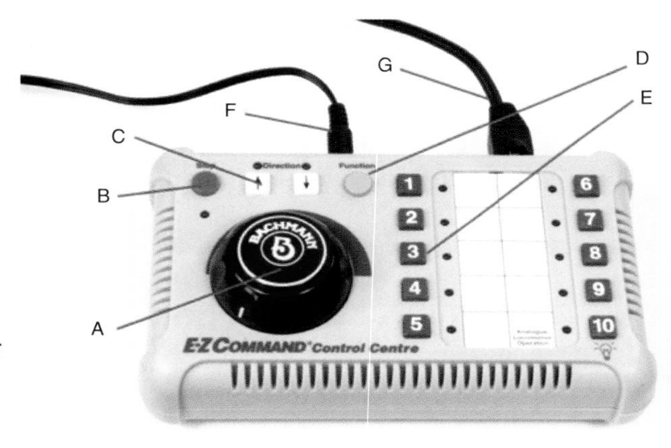


*Entry-level sets include the Bachmann E-Z Command system, which allows operation of nine digital locomotives. The controls on this combined command station/booster and throttle are very simple to use, which should be set against limited functionality. The train speed is controlled with a rotary knob (A), whilst train direction is controlled through simple forward and reverse buttons (C). Functions are controlled by pressing the function button (D) and then a locomotive address button (E). An emergency stop button is provided at (B) and all the connections required for the track bus and power supply are located in the rear of the unit (F and G).*

waveform to controlled dc current for the motor. When you consider that sound decoders are fitted with an amplifier and enough memory for sound too, not to mention the development of two-way communication in advanced decoders such as the Lenz Gold, you can see why my respect for these small circuit boards is immense. They are the locomotive control cabinets, taking the instructions from the cab or throttle and turning them into motion, speed, direction, lighting and sound. Just like the real thing.

### DCC Systems

DCC systems are usually combined with the booster and command station together in one box, called a 'base station'. The base station has connection ports for power supplies, the power bus, the throttle bus and other devices. Some base stations only work with hand-held throttles connected to the base station via the throttle bus and connection ports that can be built into the fascia of the layout; they have no throttle controls built in. Those base stations with a set of controls are called 'consoles'.

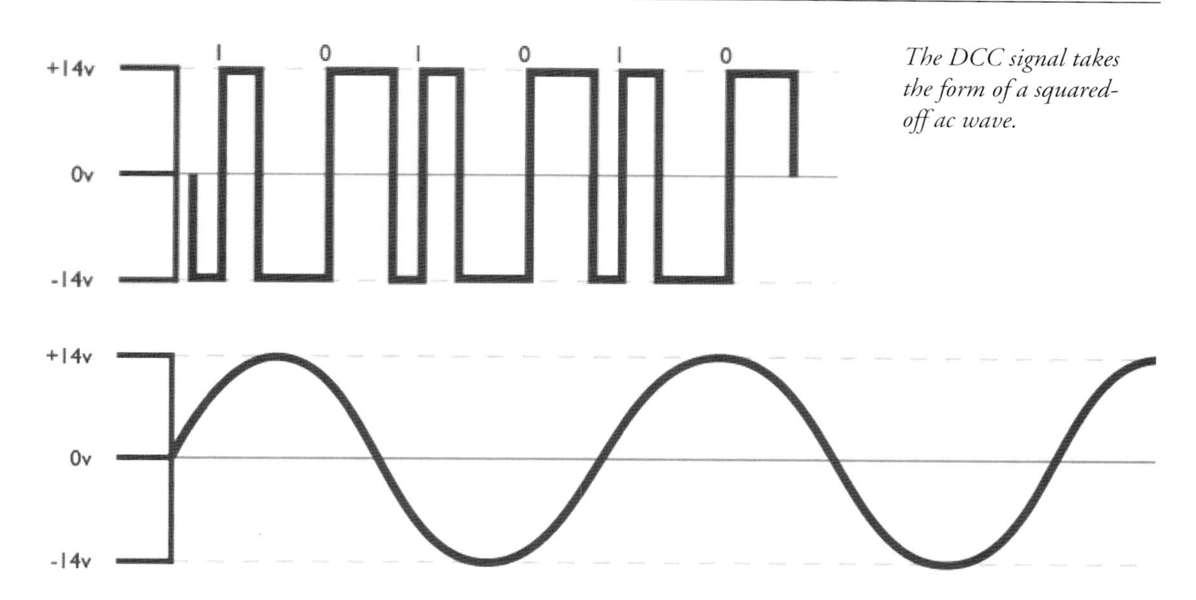


*The DCC signal takes the form of a squared-off ac wave.*

*Typical alternating current wave.*

## DCC SIGNALS

Digital packets are sent from the command station to the locomotive decoder via the power bus and the rails. There is plenty of opportunity for dirt and electrical noise to interfere with data transmission and the booster's job is to amplify the power to give it the best chance of working. Whilst knowledge of what form a digital packet or wave may take is of interest, it is not essential to successfully using DCC to control a layout. I have included this information for readers who may be curious.

The waveform transmitted by the booster is an alternating-current waveform that has a frequency of about 8,000Hz at 14 to 16 volts. This signal is similar to, but not the same as, conventional alternating current used in domestic applications, since the waveform is square rather than sinusoidal in shape. It has the power to ensure that the data packets are successfully transmitted through the layout to each locomotive decoder.

You will notice that the signal is not even in shape but is made up of a sequence of pulses, either long or short. A short pulse represents the binary digit '0'. A long pulse is the binary digit '1'. A group of eight pulses is a single byte.

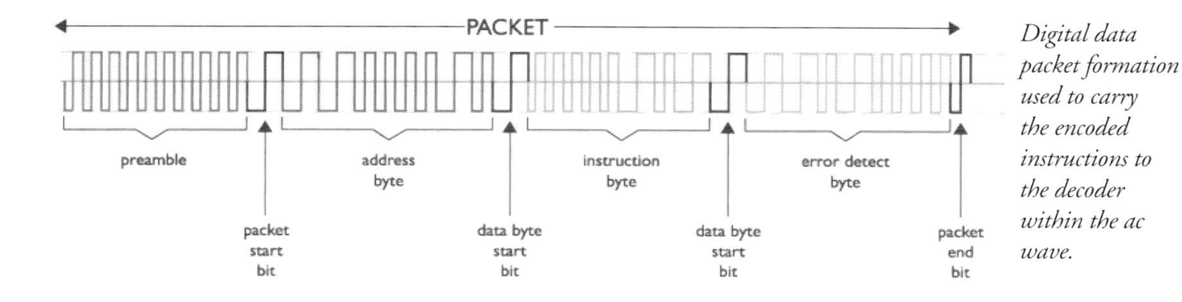


*Digital data packet formation used to carry the encoded instructions to the decoder within the ac wave.*

A data packet is made up of four discrete sections. The first section is a preamble of short pulses made up of '1' bits followed by an address byte, an instruction byte and an error check byte. Each byte has a starter '0' bit except the last one, which has a packet end bit. The packet structure is further described in NMRA Standard S 9.2 'Communication Standards for Digital Command Control, All Scales'. If you wish to research this area further, visit the NMRA Standards and RP website at http://www.nmra.org/standards/consist.html#standards.

## POINTS TO REMEMBER WHEN SETTING UP YOUR SYSTEM

- The power in the track is alternating current with a square waveform, which is very different to alternating current in domestic electricity supplies.
- DCC track power is on at all times.
- The voltage of DCC track power does not vary and does not change when the speed or direction control of a locomotive is altered.
- The polarity of electrical power does not control locomotive direction.
- Decoders control power to the locomotive motor, which means that locomotives can be stationary even when track power is constantly present. This is how lighting systems and digital sound systems continue to operate even when the locomotive is stationary.
- A locomotive will continue to obey the last instruction the decoder has received from the command station, even when the throttle has been used to select a different locomotive or has been unplugged from the system.

## THE IMPORTANCE OF INSTRUCTION MANUALS

It is a well-known saying in model railways that modellers never read the instructions. In the case of DCC, reading the instruction manual is pretty essential because it contains a wealth of invaluable information on how to get the best

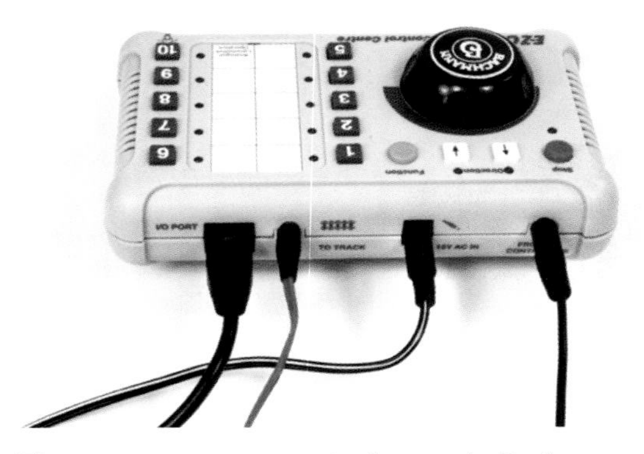

*Three connections are required to get the Bachmann E-Z Command up and running. This is basically the same for all DCC systems: a power supply from a transformer, the track bus wires that supply power to the tracks and a socket or two for the connection of walk-round throttles or companion throttles.*

out of your system. It is also worth noting that the connection of a wire from a high-powered transformer to the wrong port at the rear of your command station could result in serious amounts of smoke issuing from the seams of the equipment case and, as anyone who works in electronics will tell you, once smoke comes out, it is impossible to return it.

As intimidating as instruction manuals may appear, those supplied with decoders should also be read because they contain invaluable information on how the individual settings within the decoder are changed. This process is called programming, although in fairness you are not actually programming the decoder, merely making choices between different parameters to either improve the performance of a model or to activate or deactivate particular functions. 'Programming' is not at all difficult and some modern decoders are so beautifully set up to operate with contemporary ready-to-run model railway equipment that the only thing you have to do to them is allocate a unique address. Learning how to programme your decoders is a useful skill, however, because you will soon discover there are some interesting features that can be accessed including minimum and maximum speeds, starting voltages, lighting

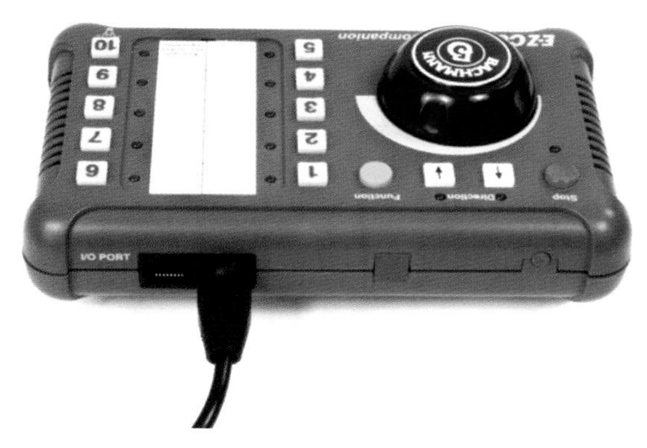


*A companion throttle for use with a base station. There is only one connection to be made and that is to the system's throttle bus so the throttles can communicate with the base station.*

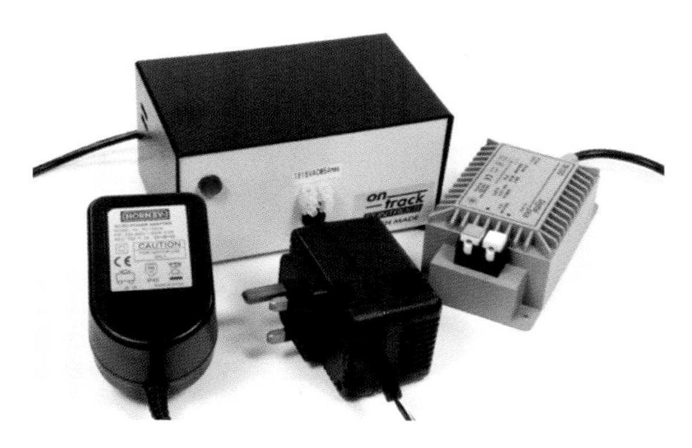

*A suitable power supply is required and not all DCC systems are bundled with one. The one on the left is for the Gaugemaster/MRC Prodigy system and is supplied with the set. The one on the right is supplied by Lenz and must be purchased separately. You are not restricted to the DCC manufacturers' transformers and the large cased transformer to the rear of this photograph is produced by an independent manufacturer.*

effects and the ability to allocate functions to particular control buttons on the throttle to enhance the operating experience. Once you have set your digital system up, store the instruction manuals in a safe place.

*Once your layout is set up for DCC operation, you can connect any 'DCC compatible' or 'NMRA compliant' system to the layout power bus. This might not be the case with regard to the throttle bus because that architecture is not covered by the NMRA Standards and RPs and not standardized between all systems. None the less, my exhibition layout is shown being connected to a ZTC-511, which is as straightforward as connecting two wires. The layout was used to test all of the DCC compatible and NMRA compliant systems featured in this book.*

## NOW YOU ARE FAMILIAR WITH YOUR SYSTEM ...

You will want to explore the system's potential and do some projects to enjoy all those benefits we mentioned in the last chapter. Before starting on the layout and installing decoders, however, there is some preparation to be done. Gathering suitable tools together, some of which you may never have considered as essential to your workbench before, is important and you may wish to create a separate tool kit for DCC work. Tools you should purchase include a wire crimping tool, quality wire strippers, wire snips, a solder fume extractor unit and a soldering iron dedicated to electrical work. Always buy the best tools you can afford because they will repay you many times over with quality performance.

There will also be some pieces of equipment you may wish to build for yourself, including a decoder tester, a current measurement device and a workbench service track for programming.

The last of these is very useful because it will save you time in not having to walk to and from the layout to use the programming track located there.

Locomotive performance is important and running-in of models is discussed in Chapter 4. It is worth buying a rolling road that can be fitted to the programming track and an adjustable type such as the Bachrus Saddle system is very useful if you are modelling in EM or P4. These simple 'getting started' projects are described below.

## LOOKING AHEAD TO DCC MODELLING PROJECTS

So what sort of projects can you do? Here are some examples.

- Install working headlights and tail lights in a variety of locomotives. Running lights take all sorts of forms and include US ditch lights, high intensity headlights, marker lights, tail lights and strobes. You can fit convincing representations of oil lamps on steam locomotives, too.
- Pop an oil lamp on the rear coach of a passenger train or programme a decoder to create the modern flashing type for block freight trains.
- Experiment with interior lights in coaching stock. Switch them on and off at will. The same goes for additional lighting in locomotives: cab lights look good on detailed models and an engine room light would grace the Hornby Class 60 model.
- Install digital sound using speakers and specialized decoders that have expanded memory for sound samples and an amplifier.
- Fit a steam generator in steam locomotives. The Seuthe range of steam generators work well with decoders.
- Fit special effects lighting such as flashing warning lights on industrial locomotives and door lock warning lights on multiple units and coaching stock.
- Have a go at using advanced accessory decoders for turnout control. Many of them can be programmed with route setting and other features.
- Look at 'Asymmetrical DCC' for added control features, including red signal stops, train shuttles and other features.
- Add a personal computer through a computer interface. You can use specialized software and DCC feedback to control a layout at one extreme or to set up a signal panel with route setting at the other.
- Look into automated coupling systems that can be operated with a decoder. The popular Kadee coupling can be adapted for DCC controlled uncoupling.

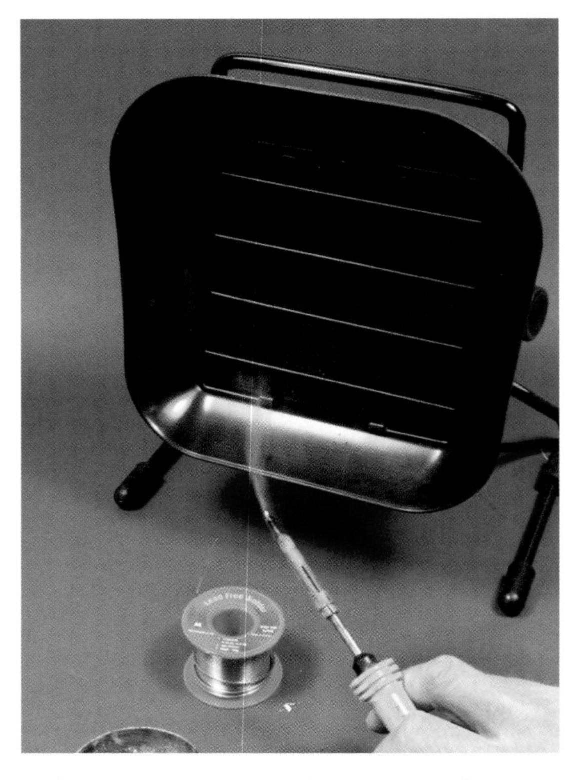

*Safety is an important consideration, especially when working with soldering irons. There are numerous projects, such as those for lighting, that will involve time spent soldering at the workbench. Buy lead-free solder and, as an added precaution, invest in a solder fume extractor like this one. Drawing away solder and flux fumes that otherwise insist on drifting into your face makes the modelling environment much more pleasant.*

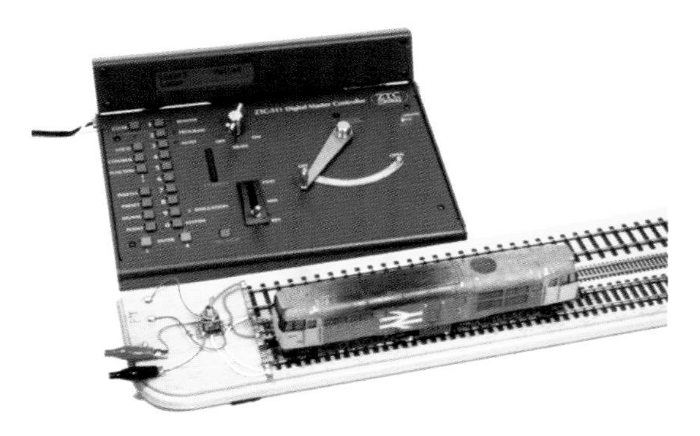

*Workbench service track for testing, programming and running in locomotives. You are not restricted to how many lengths of track can be used: several gauges of track are laid on my service track so I can work on N gauge, EM gauge and OO gauge models.*

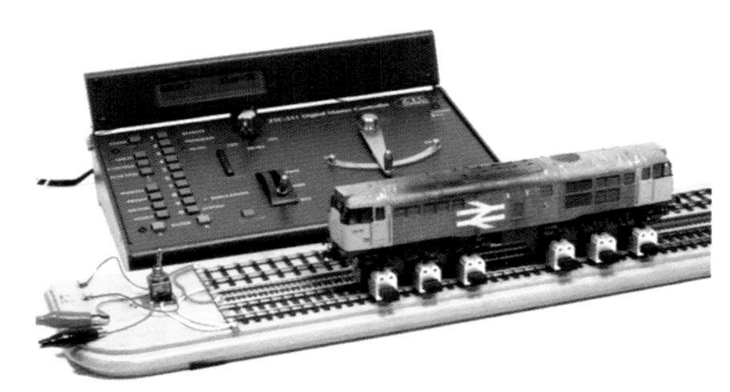

*The service track can be used with rolling road saddles such as this universal type manufactured by Bachrus. This makes running-in of locomotives very easy to do and also provides a stationary test rig for listening to the whole range of digital sounds programmed in a particular sound decoder after installation in a locomotive.*

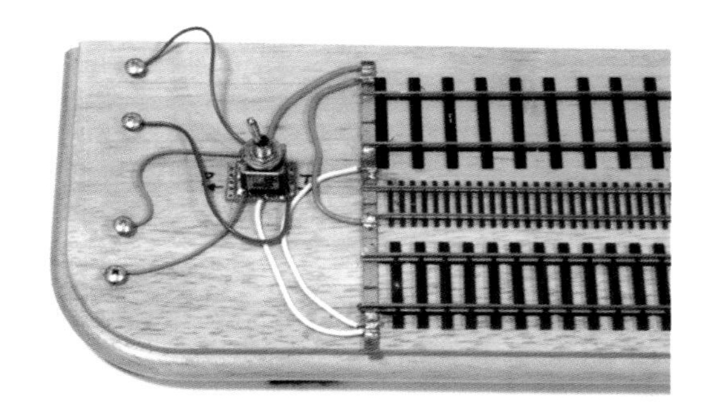

*The service track is wired up with terminals for connection to the programming track output and the main power bus output from the base station controlled by a double pole, double throw switch to facilitate an easy switch-over from service track power and the main power bus supply.*

## DON'T FORGET: TAKE ONE STEP AT A TIME!

Don't suffer burn-out by trying to do everything at once. Concentrate on achieving reliable running throughout your model collection before adding complexity to the layout.

## PROJECTS TO GET YOU STARTED

### Workbench programming and service track

I constructed a simple service track on a piece of polished shelving wood using OO/HO, EM and N gauge track so that models could be tested on the workbench. The service track was fitted with terminals for a programming track input and a main track power input, switched with a double pole, double throw (DPDT) switch so I could easily switch between the two. Bachrus rolling road saddles fit the service track perfectly and it can be used to programme locomotives, run them in or simply just run them up and down to look for lighting or running problems.

You can build a similar service track with your choice of track gauge and scale – there are three running lines on my service track because I model in British EM gauge and US N scale. A OO/HO one is essential for testing new models straight from the box. All you need is a yard of your chosen trackage system, a suitable piece of wood, a DPDT switch and some equipment wire to connect all the tracks together. Crocodile clips are used to connect the cables leading from the DCC system to the terminals on the service track. Construction time is about an hour.

### Simple Decoder Tester

This simple device was designed by Colin Walker of Inverness & District MRC in response to a need for a stationary device for

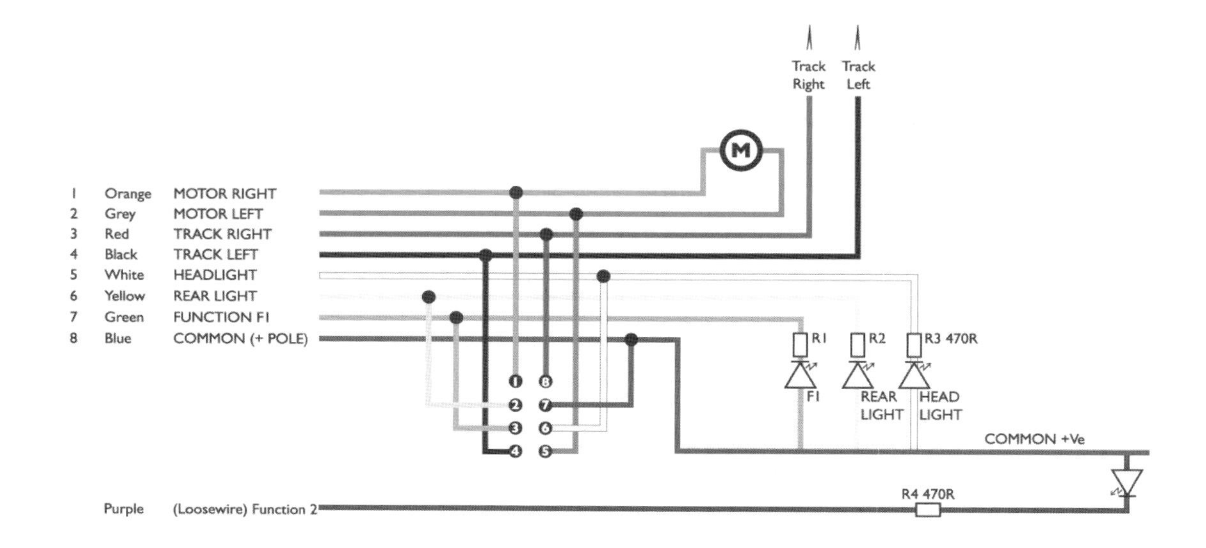


*Schematic for a simple decoder tester.*

testing and programming decoders independently of a locomotive. It provides a 'load' in the form of a motor that the command station can detect. The motor also indicates when programming commands are being accepted by the decoder – it twitches, something difficult to see if the load is provided by a resistor. It is a simple device to rig up with a handful of LEDs to test function outputs, resistors, a spare 8-pin DCC interface socket recovered from a locomotive circuit board, a screw terminal strip for 'hard wiring' decoders to it and a load motor. Construction time is about two hours.

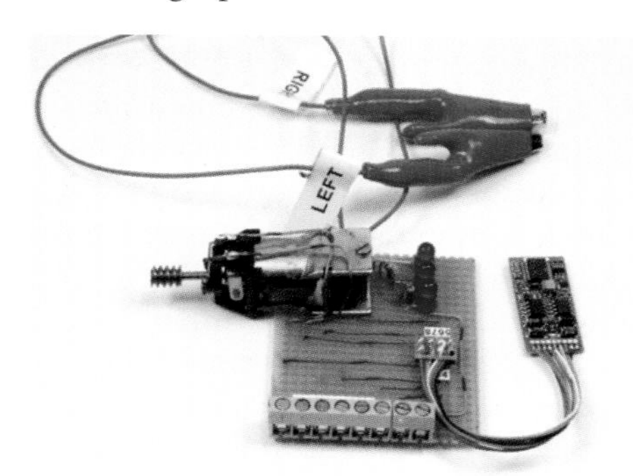


*A good project to experiment with is the construction of a simple decoder tester. This device acts as a 'stationary' locomotive so you can test your programming and even use it as a 'load' to programme function-only decoders independently of others before installing in a locomotive.*

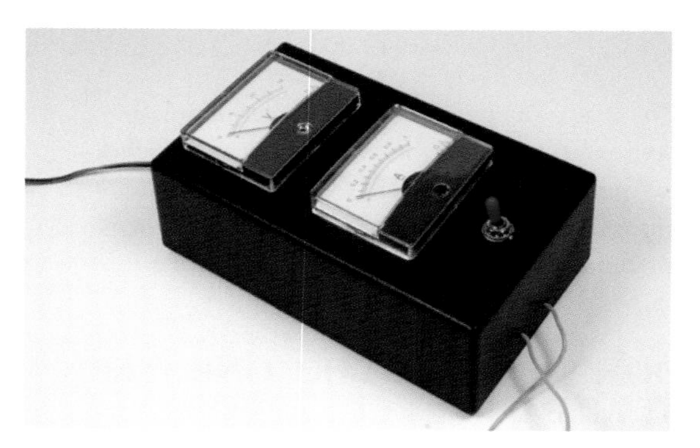

*Decoders are limited to the amount of current they can handle and the rating is stated in the decoder manuals. To be sure that your locomotives will not exceed this rating when running at speed, before choosing a decoder test them with two panel meters, one designed to measure voltage and the second to measure the amps drawn by the motor. This home-made meter box is used with the workbench service track described above.*

### Workbench Locomotive Current Consumption Tester

This device is simply used to test the performance of a locomotive before a decoder is fitted. Decoders have a maximum current rating that should not be exceeded. If a locomotive draws more than the continuous rating, you run the risk of damaging the decoder. Testing each locomotive on dc power first is one way of avoiding decoder meltdown and that can be done with a simple amps test rig. My test rig consists of two panel meters fitted to an equipment box with a DPDT (centre off) switch to reverse the current so that the meters can be used in the correct direction without having to constantly reverse the controller current. One meter is an ammeter fitted in series with the electrical circuit. The second is a voltmeter connected in parallel to the load. This also enables me to test amps versus voltage and the start voltage of a model.

CHAPTER 3

# DCC and the Layout: Wiring it up!

Getting down to the nuts and bolts, or rather the wire and solder of wiring a layout, is a task that many modellers do not necessarily relish, myself included. None the less, it is worth taking time and care over this to obtain a reliably operating layout. I focus on what is important: looking forward to the pleasure of operating the layout in a prototypical manner when it is complete rather than become embroiled in the minutiae of wiring philosophies.

When it comes to stringing wire under the baseboards, many modellers believe that they are about to 'wire the layout for DCC', as if it's something different to any other layout wiring project and that there's a great deal of mystery and black magic associated with it. Well, that's not strictly true. All you will be doing is wiring up your layout to supply power to the rails so you can get straight on with the fun stuff: operation, scenery, detailing and enjoying your models. Wiring is simply a job that has to be completed to make it all work, just as time has to be spent on baseboards to support the layout and track has to be laid to run trains. Wiring is a means to an end and not the end in itself, despite what some hardened DCC experts may wish you to believe. As always with such things in model railways, there is more than one way of going about a wiring project and this chapter suggests several methods.

*Once the baseboards are complete, the more exciting job of laying track can commence. At this point of the project there is usually a great deal of expectation, since visualizing the finished layout becomes easier and placing favourite locomotives and rolling stock on it makes it seem more of a reality. Once the track is in place, however, there is the task of installing wiring, turnout motors, connectors and a host of other bits and pieces to make it work.*

Completing the basic wiring to get something up and running can be achieved very quickly indeed. If you can wire a conventional layout for cab control with DC transformers and one or more controllers, your DCC layout will not present any challenges at all. The best advice is to take it one step at a time, starting with the wires that supply power to the rails (power bus) so you can get something up and running. It's a good idea to do that in stages, testing your installation as you go. I find it easier to work on a layout to be controlled by DCC than to complete a layout for cab control simply because there is less wiring, less soldering and less time spent under the baseboards. Some of the project baseboards were completed very quickly indeed.

I have used a new exhibition layout I was building at the time of writing to take the

*Careful planning will go a long way to help prevent wiring mistakes. As the track is test-laid upon the deck of the layout to see how the geometry works, it's worth considering how the wiring will look underneath the layout. Plan for the minimum requirement of a power bus, throttle bus and an auxiliary bus for such accessories as lighting. Depending on the type of DCC system you have purchased, you may also need to consider a feedback bus if advanced operations are your objective, together with a second set of power bus wires to power accessory decoders, separated from the track power supply by a booster or circuit breaker. Check to see if your chosen turnouts have to be modified to improve reliability: reading product manuals is very important at this stage!*

accompanying photographs to show how I installed a power bus, dropper wires to link track sections to the power bus and a 'throttle (cab) bus'. When trains were running smoothly, I installed the point motors and a separate 16V ac supply for lighting, and connected accessory decoders to the power bus to drive turnout machines (point motors) using the command station.

## PLANNING A WIRING PROJECT

It may come as a surprise that the basic principles of wiring are the same for DCC as they are for dc cab control (analogue control) and that it can be as simple as you would like it to be. There is no significant technical difference between wiring an analogue layout or one that will use a DCC control system other than that there is considerably less wiring involved and it is not mandatory to control the power in individual track sections or electrical blocks with switches (cab control or electrical blocking) to control your trains. You can have a digital layout up and running with just a power bus of two wires and a few dropper feed wires. Nothing more. The hand throttle can be plugged into the base station using the built-in socket without the need to install a throttle bus or use a console base station used with the driving controls built into it. For example, a small layout equipped with train-set trackage system, hand-operated turnouts and no signalling can be as straightforward to wire up as you could possibly imagine!

While a new layout project is the best place to start with DCC, it is possible to adapt an existing layout wired for analogue control, too, and quite simply (*see* below). This is because there is very little difference in the actual practical wiring techniques for either system. Consequently it's not surprising that very little information on how to 'wire a layout up for DCC' exists on the Internet or in print. None the less, before you make a shopping list and head off to the model shop for layout building supplies, there are a couple of things you should consider first:

- Have you tried some DCC systems first? Visit reputable suppliers to try them because there are many differences between the various systems.
- Full-blown DCC systems supply up to 5 amps to the power bus wires and track (8 to 10 amps for O gauge and larger scales). The choice of layout equipment wire is therefore important.
- You will need more than two actual wires for the power bus because each track section should be fed with current, and insulating rail joiners are used to ensure there are no short circuits caused by feeding power through the wrong end of an 'Electrofrog' or live frog turnout – in exactly the same manner as analogue control.
- A short circuit can be difficult to track down, so it's useful to provide some method of dividing up your layout into 'sub power districts' for tracing faults. This can be as simple as using connectors.
- Is the layout built to be portable (that is, it can be taken apart for transport to exhibitions) or is it a permanent home-based layout that cannot be easily dismantled?
- How many operators are likely to be working trains at any given time? It may be worth considering 'power districts' to separate working zones such as mainline and yard working. There are several ways in which you can build in power districts.
- While DCC offers the chance to remove much of the repetitive wiring and large multi-pole connectors associated with section switches located on a 'console'-type control panel, it is worth considering some places where track sections could be isolated for operational purposes, particularly in fiddle or staging yards.
- Think about the number of separate circuits you will require and plan your work accordingly. For example, my project layout has the following: power bus, throttle bus, and a 16V ac circuit for independently powering digital devices, electromagnet uncoupling devices and lighting circuits. Each will need its own colour of wire, switches and planning to determine where control switches (if used) are to be located. You may need an additional transformer for a 12V dc supply or a bridge rectifier placed in a circuit that taps a supply from an ac transformer for features such as lighting in buildings.
- Avoid taking power taps off the main transformer powering the DCC system because every tiny bit of power should be reserved for running trains. Consider how many auxiliary power supply transformers you may need. Some DCC devices, such as boosters, may need their own independent power supplies.
- How do you propose to arrange turnout control and route setting? Will you locate turnout control on a separate panel or operate turnouts with accessory decoders? That will determine how much more wiring is required beyond the power bus. It's more expensive to use accessory decoders but there is considerably less wiring involved and some of that cost is mitigated by not buying switches. A conventional control panel is marginally cheaper but more laborious to rig up properly and also ties you to a 'console'-type control method. Console control panels are not an issue if the layout is small or the operating sequence calls for an operator to be allocated

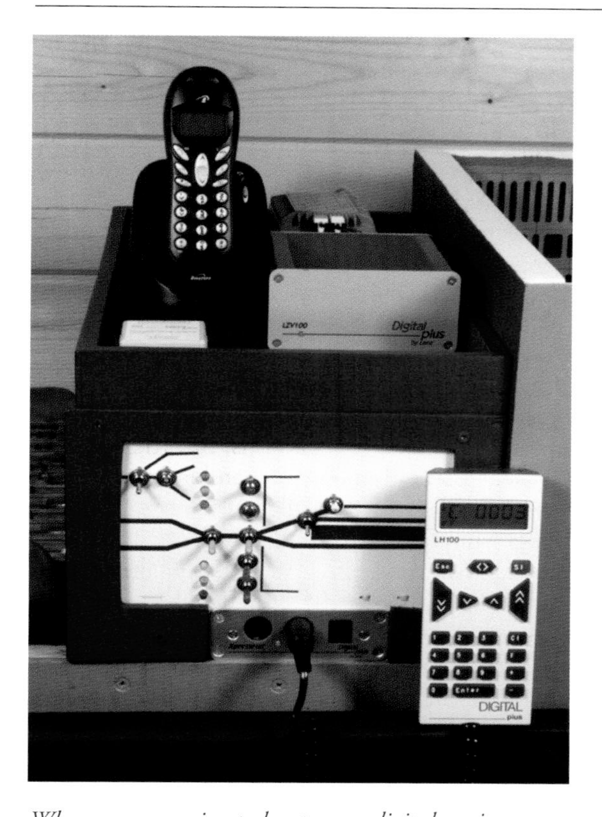


*Where are you going to locate your digital equipment on the layout? Some equipment can be concealed beneath the baseboards while others, such as the Digitrax DCS100 command station/booster, have to be accessible because switches are located on the front fascia. This is the arrangement on one of my exhibition layouts.*

the job of signalman or dispatcher. Roaming operation over large layouts is best rigged for local control of turnouts (manual control in yards) or driver operated from the throttle (there are many prototypes for that kind of route setting operation on heavy rail as well as light rail operations).

- Finally, do you need to install everything straightaway or can you do the project in stages, so at least to have things up and running? Planning your project in stages can spread the cost and provide several easily attainable objectives.

*Decide on equipment such as turnout motors and whether you wish to control them from a panel or through the use of an accessory (points) decoder – or both! It's important to read the manuals because some accessory decoders will not operate with motorized turnout machines such as the Tortoise motors. Some accessory decoders will permit a switch or push-button override, others will not. The Hornby stationary decoder, for example, is only equipped to operate solenoid-type turnout machines.*

## WIRING MYTH AND RAIL RESISTANCE

One of the common myths about DCC is that only two wires are needed. While this is technically true, there is a need to ensure that current supply and digital signal strength is consistent throughout the layout. Nickel silver rail is, most unfortunately, a poor conductor with a resistance around ten to twenty times greater than copper wire. Consequently, relying on the running rail to transmit power to the furthest reaches of a large layout is not a good idea, tempting though it is!

Furthermore, rail joiners are notoriously unreliable, adding to the resistance of the track, and can cause further voltage drop, degradation of the digital signal and sometimes become a source of faults after a period of time. If there is too much reliance on the rail to carry track power, the furthest reaches of a large layout soon have insufficient power to operate decoders properly. Poor power supply can prevent the all-important short-circuit detection in the

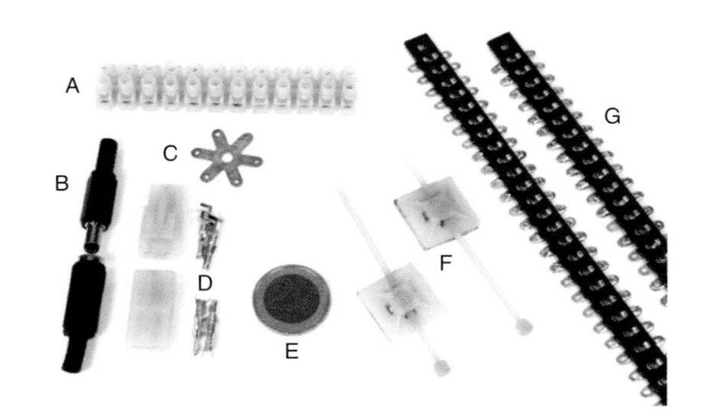


*Connectors and components of various types are very useful for portable layouts so the wiring between baseboards can be easily disconnected for transport. Furthermore, connectors such as the Tamiya type offer a quick way of disconnecting key equipment such as command stations/booster sets from a layout for use elsewhere. (A) Simple screw terminal block; (B) twin pole power connector; (C) star tag connectors for direct soldered connections; (D) Tamiya-type connectors for inter-baseboard connections; (E) coin for the 'coin test'; (F) zip ties and holders to secure wiring under the layout; (G) long tag strips.*

command station from detecting quickly enough a potentially serious problem such as a derailment that could damage the system. I personally do not rely on rail for electrical conductivity and rail joiners for reliable electrical contact.

## ESSENTIAL WIRING TECHNIQUES

### Enter the Power Bus

The answer to reliable power supply is to install a power bus consisting of two high-quality copper wires 'around' the layout, which can be connected to the track via dropper feed wires at regular intervals. The Digitrax manual and the NMRA, among others, suggest a feed to both running rails every 6–10 feet (2–3m) of running line. I personally connect every individual piece of rail so no rail joiners are expected to conduct power. The larger the layout, the greater the gauge of wire that should be used to avoid voltage drop over long wiring runs of some 40 feet (12m). The greater the power supplied by the booster, say for larger scale applications, the greater the gauge of wire that should be used. This will need to be factored in when planning the technical aspects of a layout.

### A Brief Look at Turnouts

Do you recall my reference to the urban myth of the DCC-friendly point or turnout? As stated (*see* Chapter 1), the rules governing electricity in turnouts (points) is no different for a DCC layout than it is for a dc one. In principle, there is nothing particularly onerous for the modeller to consider when laying track on a layout destined to be operated with DCC. There is no need to modify turnouts for DCC because they

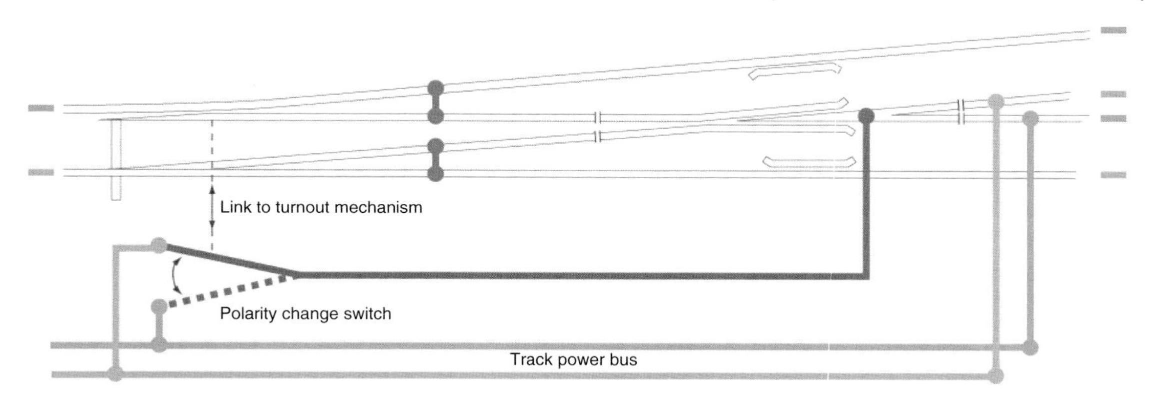


*Basic turnout wiring with polarity change switch as power bus wiring.*

will behave in exactly the same way as they would with analogue control, although some types of turnout, such as the Peco 'Electrofrog' type, do benefit from small modifications to improve their reliability, or to route power through it differently to the way the manufacturer intended. Such modifications are usually optional.

So how do turnouts work? They vary from type to type and the choice of turnout is very much a personal one dependent on the preference for a particular track system. To test every turnout for the purposes of this book would be a huge undertaking. Fortunately, most manufacturers provide details of their trackage systems and how DCC will perform with turnouts and crossings. Hornby track, for example, remains popular with modellers who wish to have a simple track geometry system to work with, durable track such as Kato Unitrak is growing in popularity, while Peco Streamline track remains a hugely popular choice.

Taking a brief look at Hornby, these turnouts automatically feed power into the track for which the turnout is set, so automatically isolating the track against which it is set. This feature, known as 'power routing', is very useful when used with analogue control because locomotives are automatically isolated in sidings when the turnout is set against them. The power routing and isolating feature can be overridden for use with DCC, and Hornby supplies simple sprung metal clips that fit between the stock rail and switch rail, making the turnout completely live no matter which way it is set.

This simple idea came about so that Hornby could claim that a layout can be powered in its entirety using DCC and relying on power transmission through the rails and rail joiners, despite the issues of electrical conductivity in large layouts. Given that Hornby has an eye on the train-set market, this is not an unreasonable option to provide because the need to establish a power bus on a table top is unlikely to find favour with the temporary Sunday afternoon layout. Hornby wanted users of its 'Select' and 'Elite' DCC controllers to be able to control the whole layout using just two wires, maintaining one of the popular beliefs about DCC and making temporary layouts possible. To Hornby's credit, it works!

More serious modellers working in N, OO and O gauge, together with various narrow gauge track systems, use more sophisticated trackage systems. Peco Streamline is probably one of the most widely used track systems throughout Europe and the USA. Peco turnouts come in two electrical formats for both N gauge and OO gauge.

'Insulfrog' means the crossing vee is insulated and there is no electrical contact between the two rails that cross each other – just like Hornby turnouts. The contact strips underneath the crossing vee across each other so there is electrical continuity beyond the turnout. Insulfrog points are equipped with power routing contact tabs on the switch rails that make contact with the stock rails in the same manner as Hornby turnouts and they can be modified with the same sprung metal clips to make them completely live, irrespective of the chosen route.

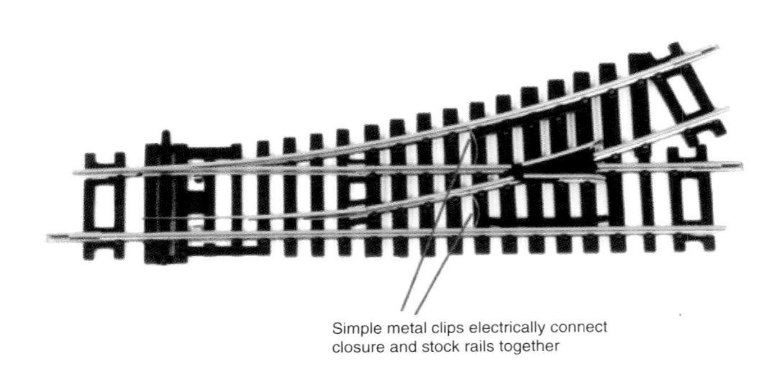


*Simple metal wire clips are used to override the power-routing nature of Hornby turnouts so they become completely live.*

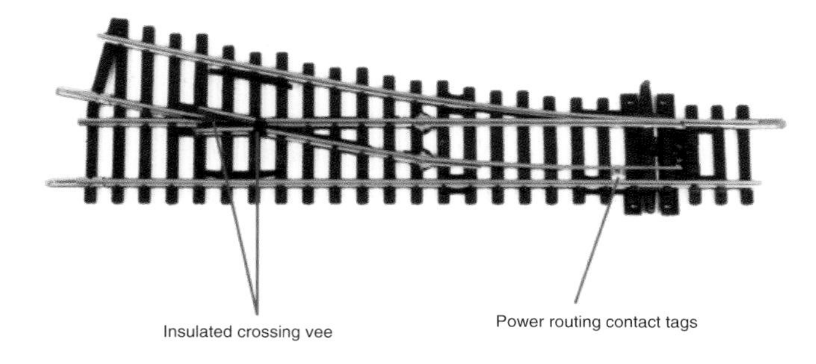


*A typical Peco Streamline 'Insulfrog' turnout.*

*A view of the underside of a Peco Insulfrog turnout shows how the crossing vee is wired up to prevent electrical contact between the closure rails. These are the easiest turnouts of all the Peco Streamline range to use on dc and DCC layouts. They can be modified with simple metal clips to make them completely live in the same manner as the Hornby turnout.*

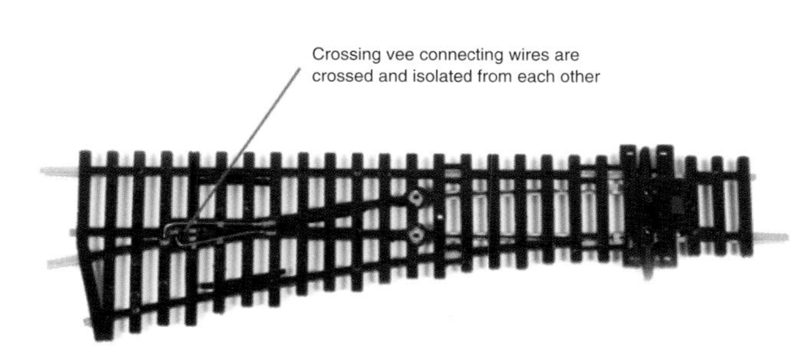


When used with a DCC control system, you can make them completely live by electrically bonding the switch rails and stock rails together. Alternatively, simply provide additional power feeds beyond the turnout so that the track remains completely live even when the turnout is switched against it. They are perhaps the simplest of turnouts to use with DCC at the expense of an insulated plastic crossing vee, which may cause slight interruption in running with small shunting locomotives, especially when operated at low speed.

Peco 'Electrofrog' turnouts are fitted with a crossing vee that is composed of metal rail and the two rails crossing are not insulated from each other in the same manner as Insulfrog points. The polarity in the crossing vee changes when the turnout is changed. However, as supplied, they are power-routing turnouts using the same type of metal tab contact on the switch rails as

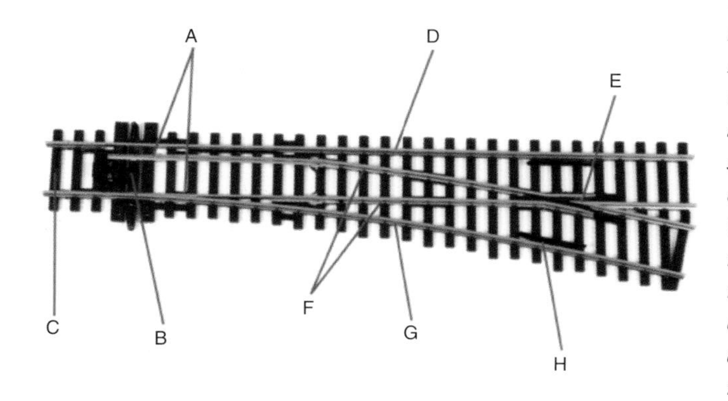


*The various rails that make up a turnout have been identified on this Peco Streamline Electrofrog turnout as a point of reference when reading this chapter. The switch rails are located at (A) and are separated by the stretcher bar (B). The switch rails make electrical contact with the stock rails (D and G) via metal tags, making this type of turnout 'power routing'. The toe is at (C), the crossing vee is at (E), which is composed of wing rails and crossing nose, and the closure rails are seen at (F). The check rail is identified on one side at (H).*

described above for Peco Insulfrog turnouts, and a polarity change switch for the crossing vee is not required.

As a result, the crossing vee is always live, making Electrofrog turnouts technically different from Insulfrog turnouts and some train set turnouts such as those produced by Hornby. Power should only be fed into them from the toe or switch end and never from the diverging end via the crossing vee, or a short circuit will result when the point is set against the road with the power feed. This is particularly important when using them to create cross-over formations on double track and on multiple track lines where power can be fed into turnouts from both ends when set for the cross-over, resulting in a short circuit. This electrical feature calls for a little more care during installation but is offset by the greater durability of the metal crossing vee over a plastic one.

A short circuit is simply avoided by fitting insulated rail joiners to the two rails leading away from the crossing vee. Cross-over formations should have both running rails isolated in this way and power feed from the toe of each turnout. When using them on a DCC layout the same rules apply: locate the power feed at the toe of the turnout and use insulating rail joiners to ensure that short circuit does not occur through the crossing vee from a power supply beyond the turnout. The crossing vee rails must be insulated in all cases, because to maintain a completely live layout to gain the benefits of DCC each length of track must be connected via dropper wires to the power bus. Unfortunately the simple trick with metal connecting clips suggested by Hornby will not work with Electrofrog turnouts.

## IMPROVING PECO TURNOUTS FOR BOTH DC AND DCC OPERATION

There is a weakness with Peco turnouts related to the electrical tags that make contact between the switch rail and the stock rail, resulting in power routing when the turnout is changed. Eventually the metal tags wear out, especially when dealing with the higher current associated with DCC. Furthermore, the way that Electrofrog turnouts are powered means that both switch rails carry the same polarity of current, where ideally you need to have the same polarity in the stock rail and adjacent switch rail. An out of gauge wheel set can bridge the gap between the opposite switch rail and stock rail causing a brief short circuit. That is not critical with analogue control but the sensitive command station will register it instantly and shut power off.

There is a simple method of adapting the popular Peco Electrofrog turnout for more reliable operation, which can be applied to analogue control, too. It involves the electrical bonding of the switch rails and stock rails together with a method of changing the polarity of the current in the crossing vee with a switch. The modification removes the power routing feature that simultaneously acted as the polarity change switch for the crossing vee in the Electrofrog version. The resulting modifications mirror the electrical set-up used for hand-built points.

The two switch rails are isolated from the crossing vee by cutting two isolating gaps in them. The switch rails are electrically bonded to the stock rails using short pieces of wire directly soldered to the rails. The crossing vee is then fed with power via a polarity change switch, which is physically connected to the turnout motor or incorporated as part of the turnout motor so that the polarity switches when the turnout is changed. All the other rules regarding the wiring of Electrofrog points still apply, including the use of insulating rail joiners at the crossing vee end of the turnout to prevent short circuits.

An exception in this family of turnouts is the Code 83 line of track and turnouts produced for HO scale US outline modelling. The turnouts are built with gaps in the closure rails, equipped with wire connections. Changes to the way the turnout routes power is easier to make – the only thing you have to do is disconnect the wires across the insulating gaps in those rails. No need to fetch the cutting disc from the tool box!

## PECO ELECTROFROG TURNOUT MODIFICATION

Peco Streamline track is one of the most popular trackage systems used in Europe and the USA. The turnouts benefit from a reliability modification described in the following sequence of photographs. Such reliability modifications to turnout have to be planned into the layout at the construction stage. Removing previously installed turnouts that have been ballasted will probably prove to be onerous and could damage adjacent scenery, which may require further effort to repair.

*An out-of-gauge wheel set can bridge the gap between the opposite switch rail and stock rail of a Peco Electrofrog turnout, causing a brief short circuit that will be detected by the command station. Changing the way that electrical current flows through the turnout will eliminate this problem.*

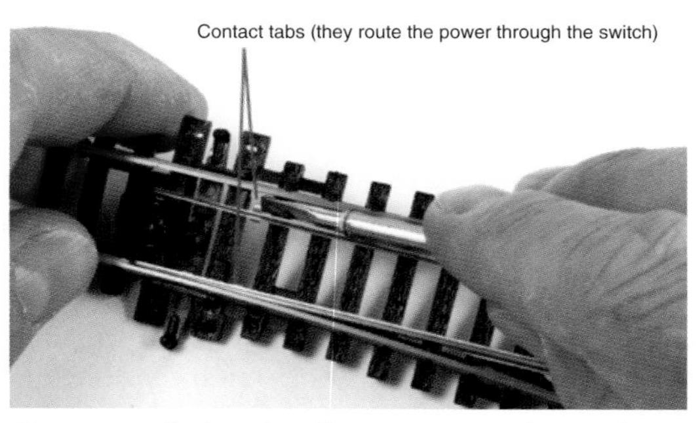


*Commence by locating the contact tags that make the electrical connection between the switch rails and the stock rails. Bend them into a position so they can be easily snipped off.*

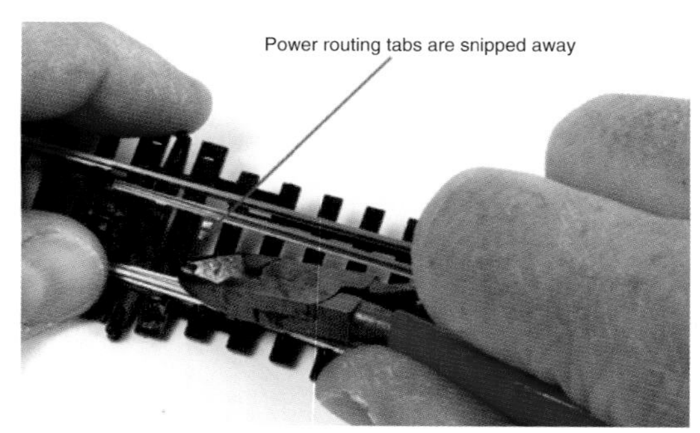


*Sharp snips are required to remove the contact tags as cleanly as is possible without distorting the switch rails.*

*Both closure rails are gapped using a carborundum cutting disc, effectively isolating the switch rails from the wing rails and crossing vee. Isolating the switch rails in such a manner removes the risk of short circuits between switch rail and stock rail, as illustrated in the first photograph of this sequence (opposite).*

*At this stage of the modification process, there is no electrical contact between the stock rails and the closure rails. In effect, the switch rails are completely isolated from any other part of the turnout. Part of the plastic webbing on the underside of the turnout is cut away so that electrical connections can be made between the switch rails and the stock rails.*

*That completes the reliability modification. On the surface there is very little to be seen other than two isolation gaps and two pieces of wire providing power to the switch rails. In effect the switch rails now share the same polarity, since they are adjacent stock rails and the turnout is no longer capable of routing the power. A polarity change switch is now required to ensure that the correct polarity of current is fed to the crossing vee for any given position of the switch.*

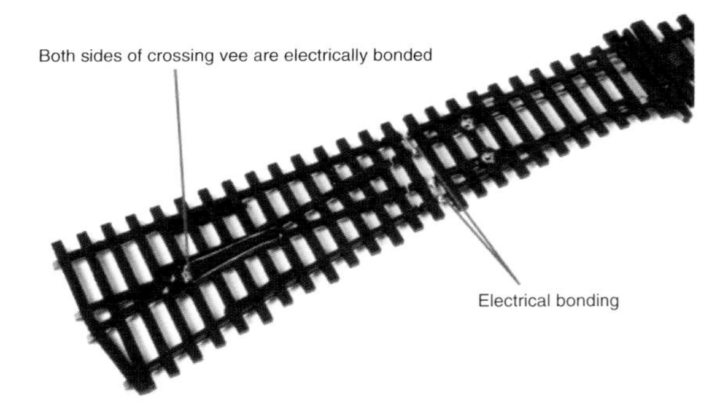


*This picture shows the new electrical connections between stock rails and switch rails.*

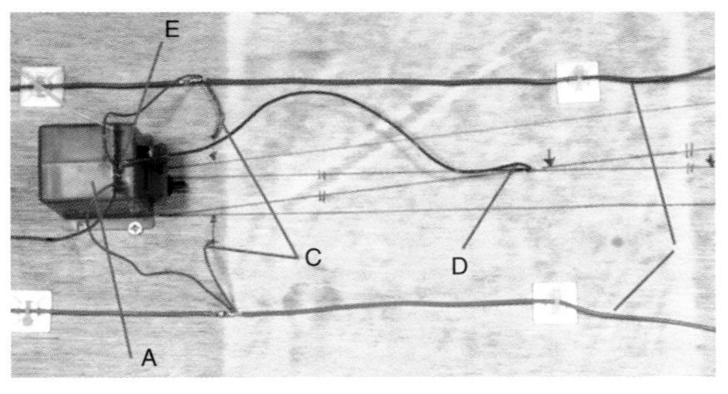


*A small demonstration panel shows how the modified Electrofrog point is wired up to a DCC power bus and using the polarity change switch in a Circuitron Tortoise turnout machine. The Tortoise is shown at (A), the toe of the turnout. The red and green power bus wires are identified at (B). Two feeder dropper wires connected to the stock rails at the toe of the turnout are seen at (C). The single electrical connection (brown wire) between the crossing vee hands of the common terminal of the polarity change switch is identified at (D). The remaining two poles of the polarity change switch are fed with current from the power bus. Care is needed to wire the connections to the polarity change switch the correct way round to prevent short circuits when trains negotiate the turnout, particularly the electrical section involving the crossing vee.*

## DEALING WITH HAND-BUILT TURNOUTS

Hand-built turnouts favoured by fine-scale modellers are almost invariably constructed with live crossing vees and the rails soldered to copper laminated sleepers or PCB strips, although fine-scale plastic turnout kits are also available that have the same electrical characteristics. This means that some precautions have to be taken to avoid short circuits and to ensure smooth operation. They are rarely, if ever, built to be power-routing in design and power is fed from stock rails to switch rails by permanent electrical connections. (Modellers keen to research the construction of turnouts further should refer to the manuals supplied to members of the EM Gauge Society and Scalefour Society.)

Fortunately for the DCC user, this type of turnout is as easy to use on a DCC layout as on a traditional analogue layout. It requires no modification for DCC at all. Simply feed the power to the toe of the turnout (from the power bus) and insulate the rails leading to the crossing vee during track laying so that power is not fed from the other end of the turnout. The crossing vee is insulated from the switch rails too, usually by cutting two gaps in them during construction of the turnout. Power is fed to the crossing vee via the type of polarity change switch discussed above, which is either incorporated as part of the turnout motor or a separate switch with a physical linkage to the turnout motor so that polarity change is automatically made when the turnout is changed. The method of hand-built turnout construction means there is no risk of contact and a short circuit between the stock rail and the switch rail if a wheel set happens to be slightly out of gauge and bridges the gap.

When working with turnouts that have been constructed with strips of PCB as sleepers or ties, it is essential to ensure that the insulating gaps cut into each PCB sleeper are made correctly, because the slightest sliver of metal or the merest contact of copper across the sleeper will result in a short circuit and the sensitive

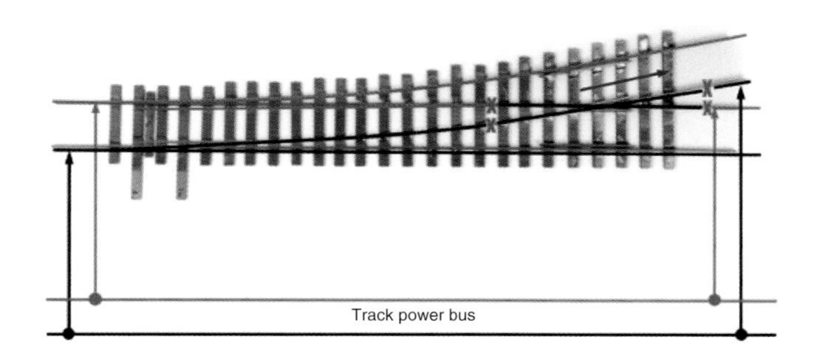


*The electrical arrangement of a typical hand-built turnout, showing the polarity in each rail when the turnout is set for the diverging route. The isolation gaps that prevent short circuits with the metal crossing vee are identified with an X.*

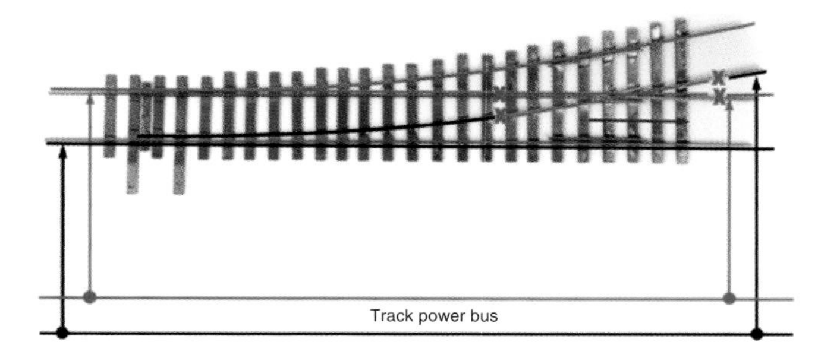


*Note how the polarity of the crossing vee changes when the turnout is set for the straight route. Such a turnout will require a switch linked to the turnout's mechanism to change the polarity of the crossing vee, details of which have been omitted for clarity.*

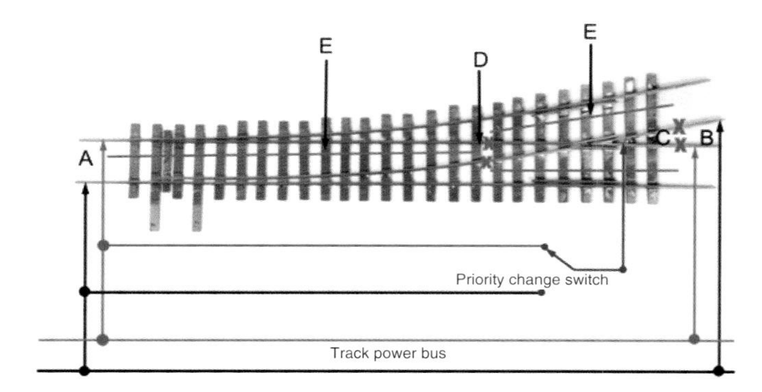


*The same turnout showing the required polarity change switch circuit that switches to the polarity of the current supplied to the crossing vee.*

short-circuit detection systems in the command station will quickly detect it and shut down the system. Also check to see that the PCB sleeper strips you are using are not copper laminated on both sides. Whilst there is no technical problem with using dual laminated PCB sleepers, there is always the chance that a short circuit may occur if the copper on the underside is not gapped in the same place as the copper on the upper side of the sleeper.

When wiring hand-built turnouts of this genre, which are very popular with fine-scale modellers in the United Kingdom, the turnout is supplied with current from the toe at (A) and that should be a feeder connection with the power bus. The electrical section beyond the turnout (B) also has separate feeds from the power bus because of the isolating gaps at X. No power should enter the crossing vee (C) from the feeds at (B). The closure and wing rails are separated by isolating gaps so that the switch rails are isolated from the crossing vee. This means that they can be electrically bonded to their adjacent stock rails so they share the same polarity. With PCB turnouts such as this, such connection can be done through the copper coating of each sleeper. This determines where the sleepers are gapped to electrically isolate the left-hand and right-hand rails from each other (E).

## ESSENTIAL TOOLS FOR LAYOUT WIRING

Most of the tools you would have to hand for building the baseboards and laying the track are also useful for completing the wiring project. The following items, none of which differ in any way from the tools required for wiring up a conventionally controlled analogue layout, should also have a place in your tool kit.

- Wire strippers and wire snips.
- Pliers.
- A variety of woodworking drill bits, particularly a 2mm drill bit.
- Cordless power drill.
- Soldering iron.
- Solder fume extractor.
- Circuit testing device, either a test meter or a simple LED and resistor circuit with croc-clips, so long as it can test for electrical continuity.
- Fibreglass pencil.
- Crimping tools for computer network cables, in case it should be necessary to extend the throttle bus using the appropriate registered-jack (RJ) connectors and cable (the choice depending on which DCC system has been chosen).

## LAYOUT WIRING

The project layout featured in this chapter shows how I wired up a layout designed to be portable. It has certain construction features so that it can be set up and dismantled numerous times and transported to exhibitions and train shows in the back of a car. The track plan allows for two separated running areas or zones: a lower yard area for shunting and switching rail wagons and an upper level that depicts a small urban station with a fairly intense passenger service. Each operating 'zone' will have its own 'sub-power district' separated by the use of a device called a 'power manager' or 'power shield', so that each is operated independently from the other, electrically speaking, even though they will share the command station/booster. Before moving on to power districts, power sub-districts and other such things, basic wiring is demonstrated by using the lower level of the project layout as a demonstration piece.

We will start by collecting tools and materials together and installing the power bus wires. You will need the following materials and should always observe the necessary standards.

- Cored solder suitable for electrical work.
- Power bus wire: two colours of a minimum grade of 24/0.2mm stranded core copper wire – ideally tinned. This grade of wire is preferred because it is capable of carrying up to 6 amps of current, which is well within the power capacity of high-end DCC systems. A higher capacity wire should be considered for larger layouts where wiring runs are considerably longer (over 40 feet/12m) and voltage drop may become a problem. For a typical mid-sized OO/HO layout, 24/0.2mm (10 or 12 AWG) wire is the minimum requirement. For large scale layouts, consider a wire capable of carrying up to 8 amps, and 10AWG or 32/0.2mm would be the minimum requirement. The logical colours to choose are red and black to match the colours of the track supply wires on a decoder harness. For the purposes of this project, and so wires could be seen clearly in the project photographs, green was substituted for black.
- Dropper wires: 1/0.7mm solid core tinned copper wire ('bell wire') is used to connect the track sections to the power bus. It is easier to solder a single core wire to the running rails than stranded wire. It's even easier if the wire is supplied 'tinned'.
- Point control wire: 16/0.2mm stranded core equipment wire.
- Snap lock connectors. One of the easiest ways of connecting the dropper wires to the power bus wires is by using automotive snap lock connectors. They simply clip on to the power bus wire without having to cut the wire or even strip the insulation from it. They are optional because there are other methods of making that connection including direct soldering or the use of terminal blocks and star tags. The disadvantage of the latter two methods is that the wire has to be cut to make the connections.
- A selection of small brass wood screws for attaching devices such as turnout motors or solenoids to the underside of the layout.
- Velcro. This material is handy in case you need to attach something temporarily to the layout. It is not absolutely essential.
- Metal eyelets for stringing the bus wires under the layout is one option. Alternatively, you could use scraps of styrene and screws for the same purpose.
- Electrical zip ties for bundling wiring runs together – just to tidy things up a little. They also find use in securing small devices to the underside of the layout.
- Connectors: 'Tamiya' connector pairs or dc power line plug and sockets capable of carrying up to 25 amps. Because the project layout has been designed to be portable, there has to be a simple method of unplugging the wires connecting each baseboard for transport. The chosen connectors have to be reliable and should only plug together one way to avoid accidental crossover of the power bus circuit. Such connectors may not be required for a permanent home-based layout, although they can be used to separate the power bus into logical sections to assist with fault finding.
- Terminal blocks are ideal for securely connecting jumper wires to the baseboard wiring.
- A large coin for circuit testing (more on that later).

Many modellers are tempted to install the power bus as a continuous ring around a circular layout. The power bus (ideally) should not be installed as a ring because of the risk of creating magnetic induction loops. The radial method, by which lengths of bus 'radiate' from the command station like the spokes of a wheel or run along the length of an end-to-end layout as a long spur, should minimize the risk and also provide a better method of fault finding. Furthermore, I avoid the use of 'Common Rail Wiring' in my layouts, preferring 'Direct Home Wiring' as recommended in the Digitrax manuals. Installing the power bus as a continuous run of wires will help prevent degradation of the digital signal and breaks should be avoided unless they are required for inter-baseboard connections on portable layouts or to introduce 'sections' to help with fault finding.

The project layout was wired up in the following order:

- The position of each piece of rail was clearly identified and a 2mm diameter hole drilled through the baseboard as close to the running rail as possible to accept a dropper feed wire. As you can imagine, there were quite a lot of holes to be drilled and carefully, to avoid damaging the track.
- Each rail was equipped with a dropper wire that was carefully soldered to the outside of the running rail. Always make sure that the wire is long enough to reach the power bus.
- With this layout designed to be portable, a set of jumper cables using the power bus wire and equipped with Tamiya-style connectors was assembled to enable connections to be made between baseboards.
- Each baseboard of the layout was turned on its side on a set of trestles to gain access to the underside. The power bus wires (red and green in the photographs) were run along the length of each baseboard, terminating at a screw terminal block located at each end of the baseboard. The colour of each bus wire was identified as left- and right-hand rail supply for connection to the command station later in the project. Identifying each bus wire and the corresponding rail will relate to which of the command station/booster output terminals the wires are connected, either terminals J and K on a Lenz system or Rail A and Rail B for Digitrax and so on. Correct identification of each wire will be important for keeping all parts of the layout in sync for the connection of further digital devices such as accessory decoders and additional power boosters (power boosters were not used on this project layout). Refer to your system manual for more information on rail sync.
- Owing to the small dimensions of the project layout it was not necessary to introduce a twist in the power bus wires. Long runs of bus wire on large layouts should be lightly twisted together to minimize the possibility of radio interference.
- The terminal blocks located at each end of the layout also provide connections for the previously assembled jumper cable sets.
- A second set of wires (orange and white) were strung along the length of the layout to provide an auxiliary power bus to supply 16V ac to additional devices such as accessory decoders or uncoupling devices. It was connected to screw terminal blocks at each end of the baseboard and provided with the same type of jumper cable assembly with Tamiya-type connectors. To avoid the risk of connectors being fitted together incorrectly when assembling the layout during preparations for an exhibition, the male and female sides of the connector sets are assembled reversed for the auxiliary power supply as a precaution.
- Dropper feed wires were connected to the power bus wires. Choice between the several ways of doing this may very much depend on the personal preferences of the individual modeller. Some prefer to use snap connectors, while others opt for a direct-soldered contact called a 'Philips Joint'.
- At this point the power supply to each section of track was tested using an old-fashioned dc controller and an analogue locomotive. Obvious short circuits can be quickly identified this way before connecting the DCC base station.

- After connecting the base station, a simple test to ensure that current is supplied consistently across the whole layout is to apply the 'coin test'. Take a large coin and place it on the track in various locations across the layout, creating a short circuit. The command station should trip out as it identifies the short circuit: failure to do so indicates that the power supply to that section of track is inadequate and additional feeder wires should be connected between the track and the power bus at that location.
- The next bus to be installed is the throttle bus, together with the associated layout 'face-plates'. You can make up your own throttle bus with the chosen network cable, terminals and crimping tool, together with the appropriate RJ connectors, all of which are usually standard items that can be purchased from your local electrical store. You can obtain a more professional appearance along the front fascia of your layout and save time by using the face-plates, throttle jack sockets and cabling supplied by the various DCC manufacturers appropriate to their throttle bus architecture. While more costly than making your own, this will have the benefit of saving time. Carefully consider where you wish to locate the face-plates along the front of your layout: it's not ideal, for example, to locate a face-plate in an aisle pinch point.
- The chosen turnout motors were installed next, together with stationary accessory decoders, some designs of which may need to tap into the auxiliary power supply. If you wish to use a traditional type of control panel for turnout operation, the additional jumper cables and multi-pin connectors will have to be assembled and the appropriate wiring installed. I must confess to preferring a traditional fascia-type panel for the control of points and signals on my smaller portable layouts. This effectively separates route control from the driving controls, just as it is on the full-size railway (except in specialized situations). However, the bundles of cable that go with such features may put me off! Larger designs may need alternative solutions, although this is very much to the taste of individual modellers. (For more about layout operation and control, *see* Chapter 7.)

## Methods of Connecting Wires to Each Other: Pros and Cons

There are tried and tested methods of making secure connections between the power bus wires and those that either link to the track sections (dropper feed wires) or stationary decoders. Beyond the need for a secure, strong connection between the wires involved there are no hard and fast rules, although ideally a method of making connections without breaking the bus wires is preferred, especially if a connection is required after the bus has been installed.

Mechanical connectors associated with the automotive industry have become popular with DCC enthusiasts in recent years. They include snap lock connectors, 'Scotchlock' and blade splice connectors. Most of them are applied to wires after installation and allow a second smaller wire or one of the same diameter to be connected to the bus without having even to strip the wires. Electrical bonding is achieved by pressing a blade down into the connector, which strips and connects both wires simultaneously. A cover is then folded over the top of the blade to prevent accidental contact with other wires. Snap lock connectors and others in that family of connectors can be used to take spurs off the bus and can be applied anywhere there is room on the bus wire. They are notably reliable and I have never experienced a failure with them.

For the budget-sensitive enthusiast, there is a simple technique that involves stripping a short length of bus wire (about 10–15mm) and directly soldering the dropper wire to it. The last 20mm of the dropper wire is stripped and the bare core wrapped around the bus wire about five times before soldering. It's quick, cheap and easy. This is an accepted connection method called a 'Philips joint'. The minimum number of turns on the bus wire must be four and the joint flooded with solder afterwards. The only drawback is the need to cover the exposed connection with electrical insulation tape to prevent accidental short circuits.

One of my exhibition layouts uses a form of tag strip connector to which wires are soldered. A variation is the 'star tag' connector, which has six tags with holes to accept the stripped ends of equipment wire. These enable reliable soldered connections to be made between different thicknesses of wire. Star tag connectors are used at the end of bus wire spurs where more than one dropper wire connection is required, as well as in a number of other locations. They are only effective if the bus wires are cut to make the connection, but such connectors can add to electrical resistance.

Screw terminal blocks are ideal for the end of lengths of power bus wire and may be located at baseboard joints where a jumper cable and connectors are to be located on portable layouts. They are not as reliable as soldered connections and so I do not use them for connecting dropper wires to the power bus wires on my layouts.

## POWER BUS WIRING

*Ideal locations for connecting dropper wires to the track are carefully identified and marked on the track bed with a black marker pen. Attention is paid to avoiding obstacles underneath the layout such as the struts used to strengthen the baseboard frames and the blocks of wood that secure the baseboard top to the frames.*

*A cordless power drill is used to drill 2mm-diameter holes through the baseboard as close to the running rails as possible. A great deal of care is taken to avoid damaging the track. The size of drill chosen is large enough to allow the track feed wires to be inserted without difficulty yet not too wide to allow ballast to escape through the baseboard during ballasting.*

*Lengths of 1/0.7mm solid core tinned copper wire are used as track feed wires. The ends are stripped and then the piece of wire is inserted into the hole with the top part folded over so that it does not drop right through and has to be retrieved from the floor!*

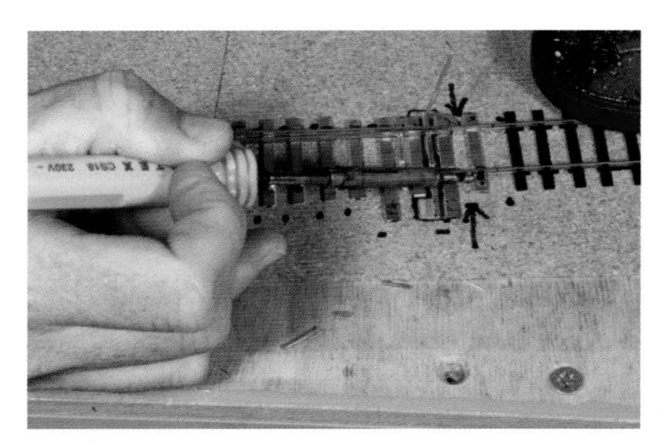

*Each track feed wire is soldered to the outside of the running rail with as little solder as possible. Particular care is taken to avoid cold joints, which could result in a wiring fault.*

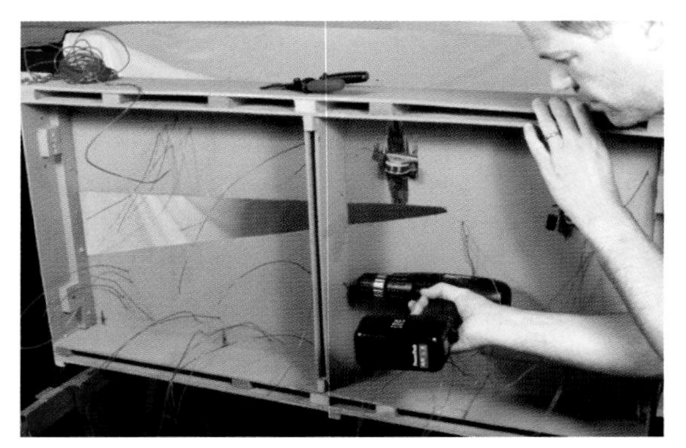

*Holes are drilled through baseboard frame cross-members to accommodate the power bus wires, among others.*

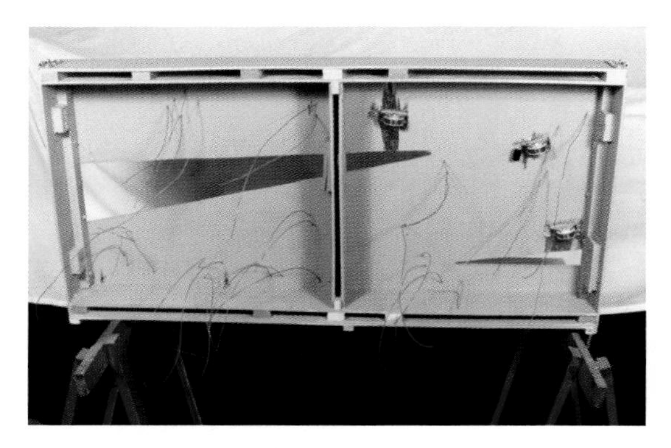

*This image of the underside of one baseboard of the project layout shows the installed track feed wires dangling beneath the baseboard. It does not really matter whether the main power bus wires or the track feeder wires are installed first, so long as sufficient wire has been left so they will meet for secure and reliable connections.*

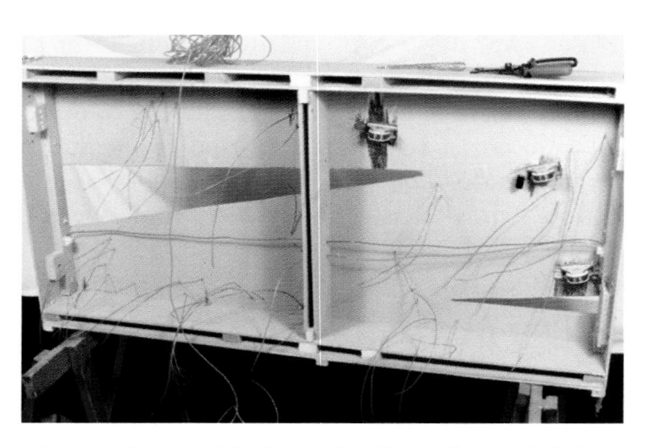

*Terminals screw blocks are fitted to either end of the baseboard and the power bus wires strung between them.*

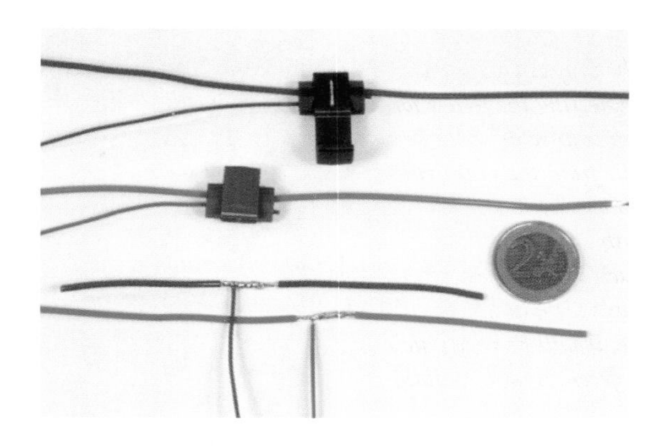

RIGHT: *The next task is to connect the track feed wires to the bus wires. There are several ways in which this can be done without breaking the continuous run of wire. Methods include the use of snap lock connectors, as seen at the top of the picture, and cheap but effective direct-soldered connections (Philips joints), as seen at the bottom.*

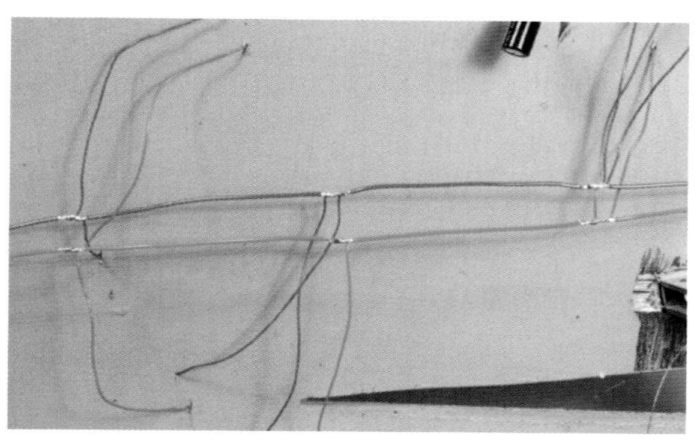

*Ideally, each of the exposed direct-soldered connections should be insulated with electrical tape once the wiring project is complete. Don't rely on being able to make the various connections staggered from each other, because in reality it doesn't always work out that way.*

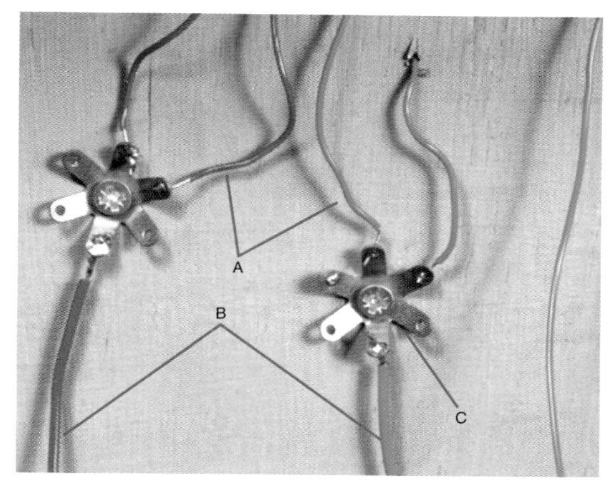


*Another reliable method of connecting the feeder wires (A) to the bus (B) is to use metal star tags (C) that allow multiple connections. This is ideal for those locations where a spur in the bus ends and is required to supply numerous track feeds.*

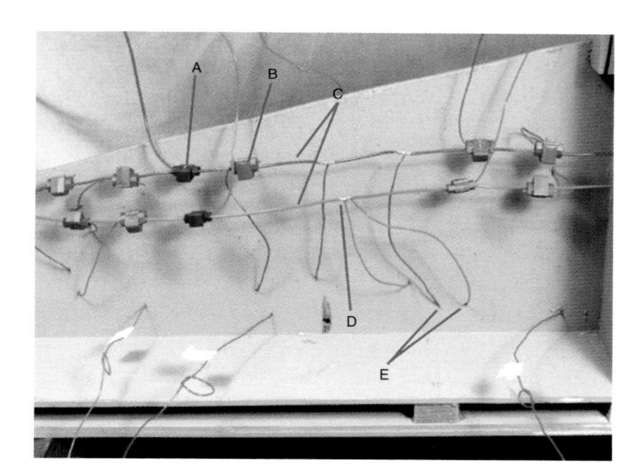


*The project layout was used to test various ways of connecting the feeder wires to the power bus (C). Snap lock connectors have been used at (A) to take a power bus spur from the main run of wire. Feeder wires (E) are shown connected with the same device at (B), although with a connector designed to accept smaller diameter wire (such connectors are colour-coded depending on the size that they will accept). Whichever method you prefer to use, it will be very much down to your personal preferences and available budget.*

*The jumper cable arrangement used to connect the baseboard wiring together.*

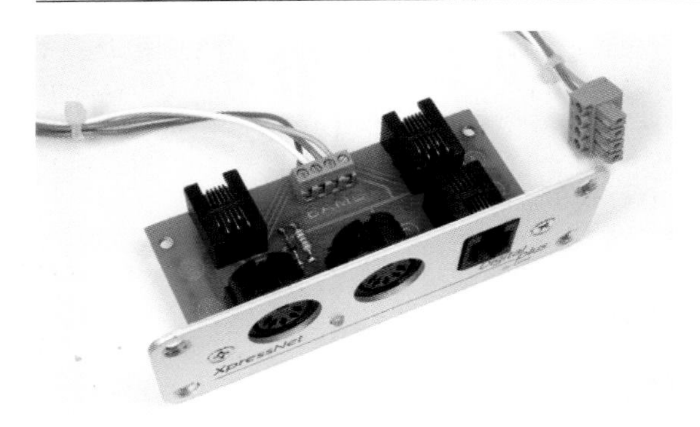


*A Lenz XpressNet throttle face-plate showing the network cable connectors at the rear. Note that a four-screw terminal block is also present, labelled BAML. These letters correspond to a screw terminal XpressNet connection in the rear of the Lenz LZV100 command station/booster, offering an alternative method of connecting devices together.*

*Throttle face-plates are designed to fit the layout fascia and a hole should be cut into the fascia to accommodate them.*

## THROTTLE, CONTROLLER OR CAB BUS

The set of wires used to connect devices such as the hand throttles and accessories to the command station may be known as the cab, throttle or controller bus; I will refer to it as the throttle bus.

As complex as this set of wires may appear, the manufacturers will have done the hard work for you since, almost without exception, they offer cables, connectors, layout fascia throttle connection face-plates and similar accessories to support their systems, all of which simply plug together. Many manufacturers include information in their instruction manuals on how to make up the interconnecting cables using commercially available parts by simply using Registered Jack (RJ) connectors and the appropriate cabling. I have used the Digitrax and Lenz systems, which can use RJ12 and 6-wire network cable, as examples to follow because it was not practicable to demonstrate all of the various types of cabling. For the record, Gaugemaster/MRC Prodigy uses 8-wire cabling and North Coast Engineering uses 4-wire cabling on its advanced system. No matter which is used, ensure you stick to the manufacturer's requirements, even if some wires in the cables appear to be unused (Lenz only uses four of the six wires in the cable).

As discussed earlier, NMRA Standards and RPs do not cover the controller side (throttle bus) of DCC systems. It is very much left to the individual system manufacturer to create its own system and the practical stuff that goes with it, including electrical cabling that connects hand throttles, controllers, boosters or other devices to the system. Digitrax refers to this network as system architecture and it is, in effect, a Local Area Network (LAN) of devices that communicate with each other. The method of communication is different between certain systems: Digitrax LocoNet is 'event driven' LAN technology while the XpressNet system is 'polled'.

It does not really matter which you choose as long as you are aware of what this means when buying equipment. In effect, throttles and similar devices from one manufacturer may not necessarily work with command stations and boosters from another. Some systems architecture use the same protocols, such as XpressNet, and in theory will communicate with each other (for example, Lenz and Roco MultiMaus). but experiment with care.

As far as wiring the layout is concerned, there are two main options for completing the throttle bus. You can either buy all of the manufacturer's components and hook them up or buy in commercially available parts and make your

*The finished result is clean, neat and professional. They are a sound way of connecting the network of throttles and devices to the command station. Two types of jack plug socket are present on this component: five-way DIN plug connections (A) for connecting controllers, a socket that will accept RJ connectors (B) (XpressNet only uses four of the six wires in the cable) and a power indicator light (C).*

own. It is possible to add simple 5-pole DIN sockets to the front fascia of your layout and wire up the throttle bus to it. A corresponding jack plug will be more than sufficient – just take care to wire it the correct way round.

You could also opt for a combination of both methods: use the commercial throttle face-plates with their sockets on the front fascia of the layout for a neat, professional appearance and make up your own connecting cables for linking them all to the command station/booster and any other devices you may wish to hook up to the network. This means you can cut cable lengths to suit your layout design and allows the custom placing of throttle face-plates to suit operational requirements. This has to be one of the easiest wiring jobs for a DCC layout and it's surprisingly simple to do.

The following tools and materials are required to complete a basic throttle bus installation:

- Throttle face-plates.
- uitable network cable for the throttle bus (4-wire for NCE, 6-wire for XpressNet and LocoNet and 8-wire for Gaugemaster/MRC).
- Suitable connectors (RJ12 connectors were used on the demo layout, but the choice of connector will be determined by the system requirements and the fit to sockets in the rear of components such as throttle face-plates, if they are being used).
- Stripping and crimping tool.
- DIN jack plugs and sockets can be used for some applications.
- Digitrax offers a testing device for checking throttle bus cables.

The use of RJ connectors for XpressNet or LocoNet is relatively simple but requires the use of a quality stripping and crimping tool. The connectors are attached to the cable wired pin 1 to pin 1 to ensure that the connections are maintained the correct way round. On a given length of cable, the connector at one end will appear to be the other way round to the other end.

Simply strip the outer cover from the network cable (only remove about 7mm) to reveal the six inner wires. Do not remove the insulation from the inner wires. Insert the connector in the crimping tool and then insert the stripped cable into the rear until the inner wires hit the end. For both LocoNet and XpressNet, the white inner wire should always be located at pin 1.

Press the crimping tool shut to close the contacts onto the wires. Release and remove the connector and wire assembly from the crimping tool. No soldering is required. The completed cables can then be used to connect the various devices and face-plates together, but avoid connecting in a loop as this will only cause problems later. Throttle bus architecture is usually designed to be 'daisy-chained' together without a loop in the system.

The connector method of connecting the various devices together plays into the hands of portable layout builders: the cables can be disconnected and carefully stored when the layout is dismantled, stored or transported to another location such as an exhibition hall.

A final but very important technical point relates to the routing of throttle bus cable. It should be kept as far apart from the power bus as possible to avoid interference. Keep the two wiring runs separate.

## MAKING AND INSTALLING THROTTLE BUS CABLES

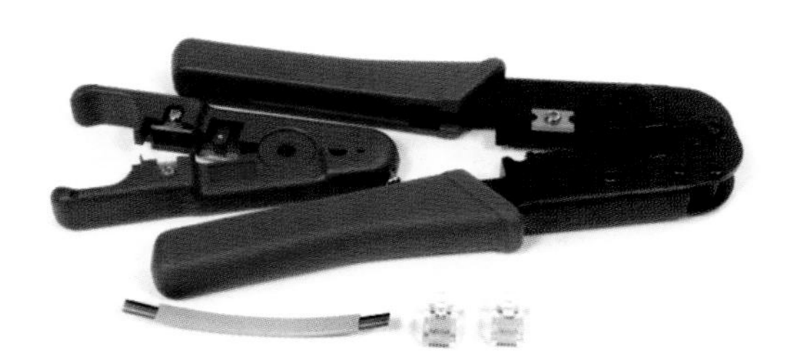

*Making up your own throttle bus cables is quite straightforward and some manufacturers' manuals will advise on how to do this. You will need crimping tools, which may or may not include specialized wire strippers, the appropriate connectors and network cable. This series of photographs shows how to make cables for the Digitrax LocoNet throttle bus. Remember to only use cable recommended by the system manufacturer. Even though some wires in the cable appear to be unused (Lenz, for example, uses only four of the six), stick with the correct number of wires since some unused wires may be reserved for use in the future as the system is developed.*

*Identify pin 1 in the connectors as indicated in this RJ connector. The cables are wired pin 1 to pin 1, also known as a one to one cable, so knowing which is which is very important. (Not all systems are wired pin 1 to pin 1, for example the Prodigy Advance system.)*

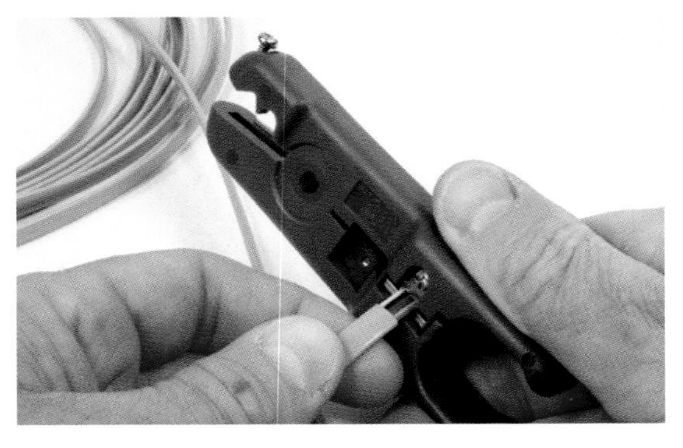

*Strip about 7mm of the outer insulation from the cable to reveal the six inner cables. The stripping blades on the crimping tool or a separate stripping tool designed for this type of cable should do the trick.*

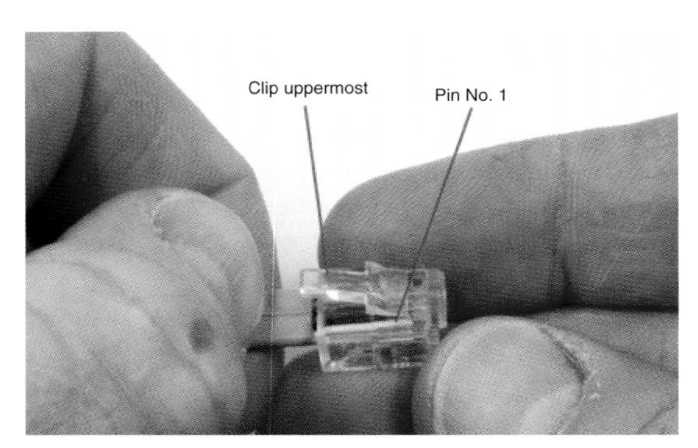


*The insulation is not stripped from the inner wires – it's not necessary. The wires are inserted into the rear of the connector, with the white wire oriented to correspond with pin 1. Note that the clip is shown uppermost with pin 1 nearest the camera.*

*The connector is crimped onto the cable. Insert the connector in the appropriate slot in the crimping tool.*

*Insert the cable the correct way round and press the crimping tool shut. If you have it right this presses the contacts into the inner wires in the correct order.*

*A very short length of cable (top) ready for use on a Digitrax LocoNet system. Note that the connectors are opposite each other at each end of the cable. This ensures pin 1 is connected through to pin 1 at the opposite end of the cable. Digitrax offers a simple tester for the throttle bus cables shown in the lower part of the picture. This is used to test home-made cables.*

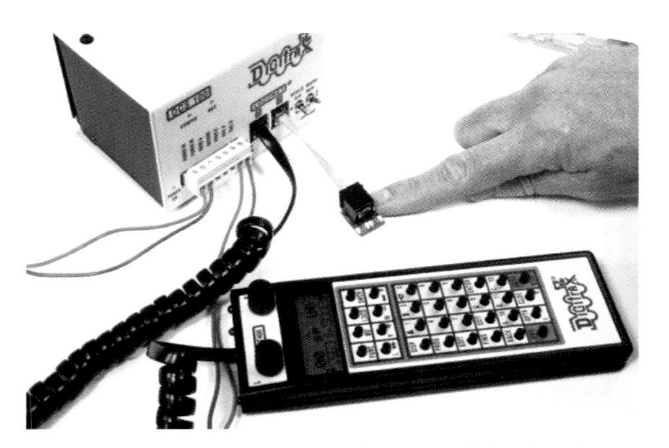

*To test the cable, it is inserted in one of the throttle bus ports in the command station. The testing device is plugged in the opposite end. If the cable has been assembled satisfactorily, all four LEDs will illuminate.*

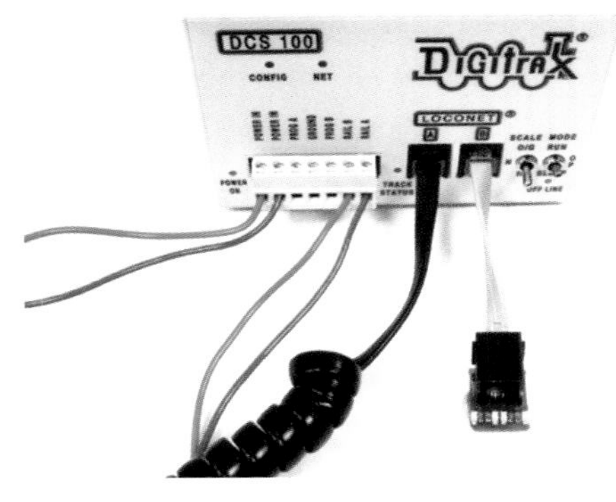


*The short demonstration cable appears to be satisfactory: all four LEDs are illuminated. If they indicate that there is a poor connection, insert each connector into the crimping tool and crimp them for a second time to see if that was the cause of the poor circuit. This may be all that is required to avoid throwing the assembly away and doing it again!*

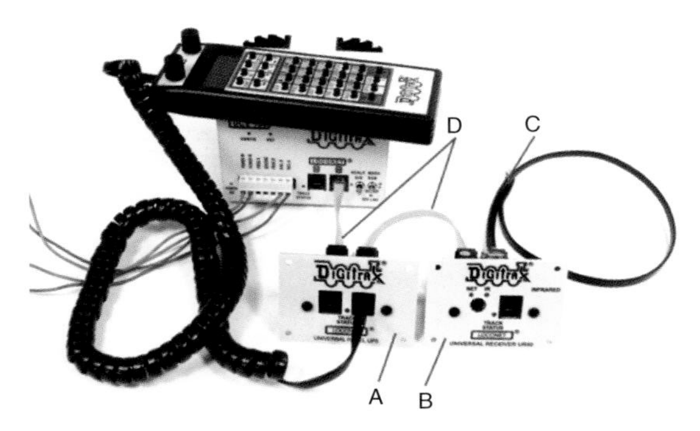


*In theory, assembling the throttle bus wiring is as simple as plugging everything together. Note here how each device and face-plate (A and B) is 'daisy chain' connected, one into the next, using lengths of throttle bus cable (D). The lengths of cable can be made up to suit the layout design and dimensions. Runs can be very long indeed and each manufacturer's manual will advise on the maximum length of throttle bus cable that can be installed (C).*

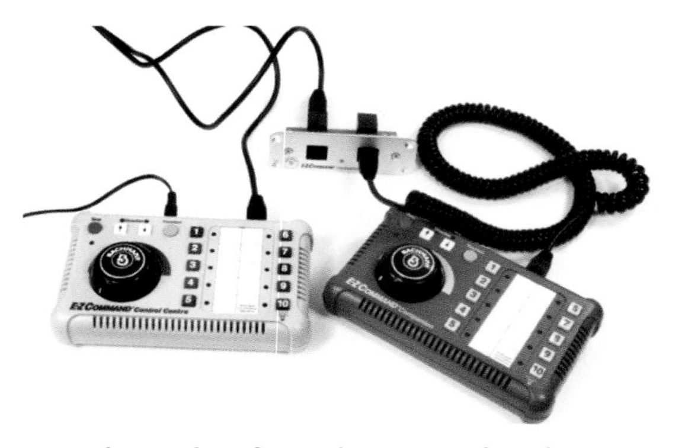

*Even the simplest of control systems, such as the very simple-to-use Bachmann E-Z Command system, uses a throttle bus to connect devices together as a network.*

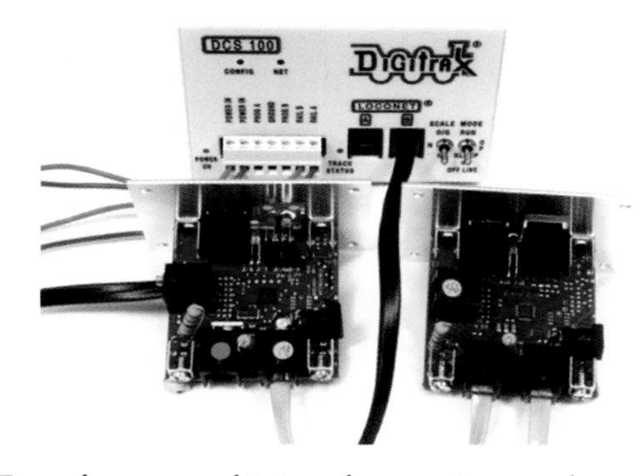


*Face-plates are sophisticated connection panels mounted in the front fascia of the layout. Additional sockets are provided in the rear for spurs off the main throttle bus run. Do not create loops in the throttle bus.*

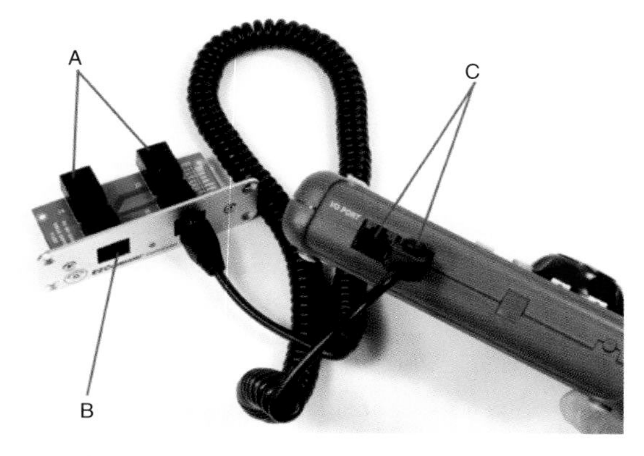


*Face-plates are an excellent way of making connections for throttles. A Bachmann E-Z Command companion throttle is here shown plugged into a face-plate as if ready to drive a train. There are sockets in the rear for connection to a command station and other face-plates (A). Sockets in the front of the face-plate (B) show connections for additional throttles. The same sockets are found in the rear of the companion throttle (C), which otherwise can only work with a command station/booster.*

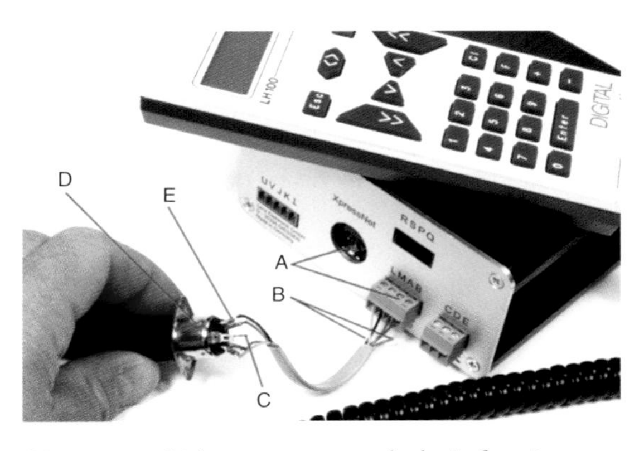


*No matter which system you use, the basic function of the throttle bus is the same: to ensure clean communication between devices such as throttles, boosters and the command station. While the exact details will vary from system to system, the use of ordinary off-the-shelf components means that they can be wired up in a variety of ways depending on the system. Ordinary wire can be used if it has a regular twist in it. This picture shows the XpressNet connections into the rear of a Lenz command station. The factory-installed connection on throttle leads and in the rear of the command station is based on the five-pin DIN plug (D). A socket is provided in the rear (A) together with screw terminals labelled LMAB for identification for the same connections. Of the six wires in the cable, the white and blue outer wires are not used (B). One pin on the DIN socket is also unused (C), while the other four cables are connected to the remaining four pins. There is advice in the Lenz manuals regarding the correct connections.*

## TURNOUT CONTROL

There are several options for turnout control and the type chosen will very much depend on the size of the layout, the way it is operated and how many operators will be online at any one time. Personal preferences are an important variable: if signalling and panels are your thing, nothing should stop you from building one. If you hate wiring, do away with the panels and consoles altogether and use accessory decoders! Here are some ideas:

- Create signalling panels just for the turnouts and signals, operated with toggle or push-button switches but with no block switching. Leave that to the throttles. Depending on the size of the layout and if you work with a dispatcher or signaller, you may wish to group all turnout and signal controls on the panel. Some US-outline modellers delight in resurrecting old Centralized Traffic Control (CTC) machines and some British modellers love installing lever frames. One thing should be considered: apply such panels to the main-line only, because yard operations should be localized.
- Combine DCC accessory decoders with localized panels and push/toggle switches. This approach helps those operating trains in the confines of a yard where constantly changing turnouts with a hand throttle can be cumbersome, even when a dedicated accessory decoder button is provided on the throttle. Select decoders carefully to see if they will support manual switch control as well as DCC commands.
- Control some or all yard and industry tracks with manual ground throws.
- Control everything with accessory decoders and throttles (or a computer interface) and eliminate switch panels altogether.

In reality, modellers are likely to select a hybrid approach to suit their particular environments. My usual preference is for all turnout and signal control to be grouped on a single panel on my small portable layouts. My plans for a larger layout include a single panel for main-line turnouts and entrances to yards together with DCC control, although my instincts are to keep signalling and route setting separate from driving controls. After all, except in some light rail applications, driver- or engineer-controlled power turnouts are rare (Montana Rail Link is one such company that uses remote-operated power turnouts for the entrance to its yards in Missoula and Helena, which are changed using radio channels, but under strict operating rules). In the same manner as full-size railways, control of yards will be localized and industrial spurs

*Choosing the right accessory decoder to suit your choice of turnout motor or solenoid is important. Twin-coil solenoids such as Seep require a different type of decoder to the Tortoise, which is a slow-motion turnout machine. Some accessory decoders can operate both.*

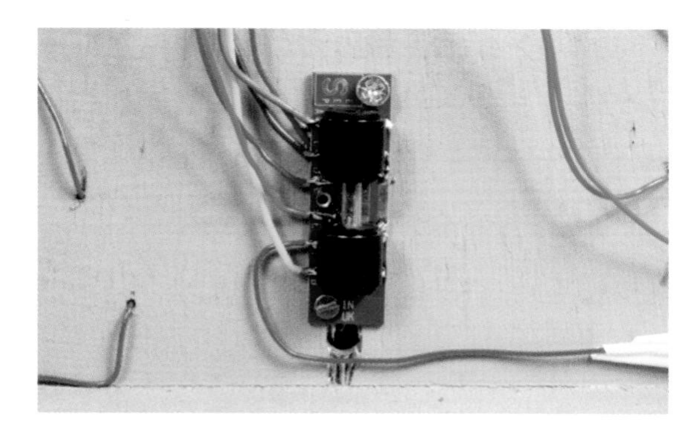

*A twin-coil solenoid fitted and wired to a layout.*

and sidings may well benefit from manual turnout control.

Most accessory decoders are designed to control twin-coil solenoid-type turnout motors such as those produced by Peco, Atlas or Seep. They require a small burst of current to throw (usually snap) the turnout over and that's it. Some twin-coil solenoids are self-latching, which means they will keep the turnout locked over after it has been changed.

Slow-motion motorized turnout machines are becoming hugely popular because their action is smoother, slower and more realistic. Such motors include the Tortoise, Fulgarex and Lemaco, which are powered by a motor and gear train. The Tortoise is a 'stall motor', which means that it uses a small amount of current at all times. This will not harm them as long as it does not exceed 12V dc (at 12V about 10mA of current is used). The typical stationary decoder will not provide power after the initial burst to throw the turnout, and latching of points without centre-over spring mechanisms may not be easy to achieve. So which type of turnout machine is likely to be your preference: solenoid or slow motion? That will determine your choice of accessory decoder type.

All accessory decoders will work off any DCC system – in theory. We have returned to the power bus side of the layout and the principles of NMRA Standards and RPs, which means that stationary decoders have to comply, just in the same manner as mobile locomotive decoders. You should be able to use a mix of stationary decoders, if it suits your needs, in the same manner as mixing locomotive decoders on one layout. Sometimes there are compatibility issues and you should be alert to the possibility.

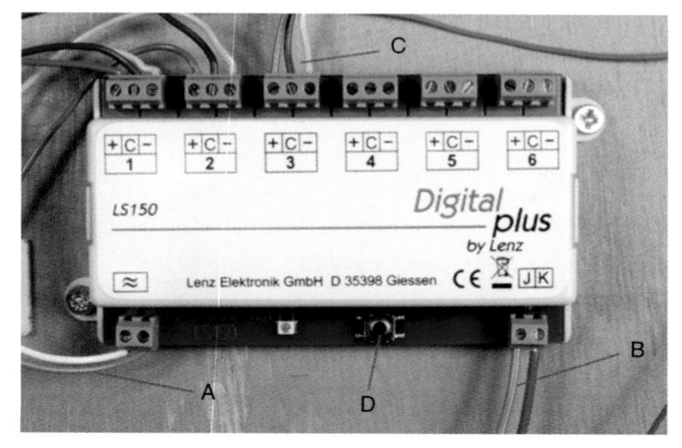


*The Lenz LS150, an ideal accessory decoder for twin-coil solenoids. It has ports for an independent power supply (A), ports for connection to the power bus (B) and six turnout control outputs (C) with three terminals each for wiring to twin-coil solenoids. The decoder is programmed using a button (D) and by observing the actions of the adjacent LED.*

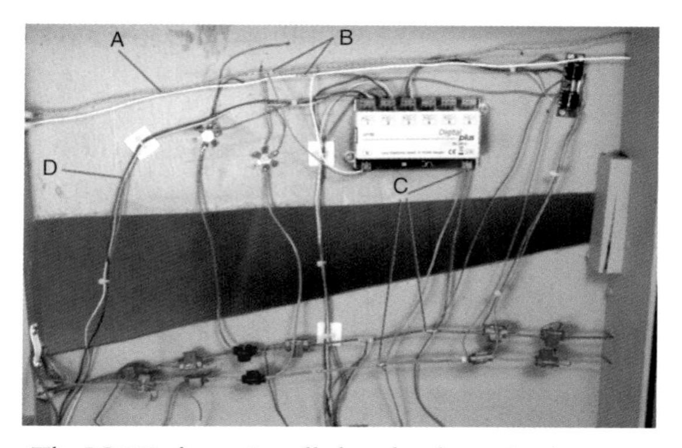


*The LS150 shown installed under the project layout with Seep twin-coil solenoids: (A) independent 16V ac power supply; (B) connections from the power supply to the LS150 decoder; (C) connection to the power bus so digital signals can be transmitted to the decoder; (D) three wires connect the twin-coil solenoids to the LS150.*

Let's take a look at decoder types:

- Some decoders provide a shot of current to throw twin-coil turnout motors. The pulse of power has a set programmed duration that can be changed in most cases. A good example of this is the Lenz LS150, which has six outputs (controls up to six turnouts). As a separate note, the LS150 will provide a sufficiently long pulse of power to throw a slow motion motor, but if there is no internal latching in the motor, the turnout switch rail may ease away from the stock rails once the power pulse ceases.
- There are decoders that supply constant power for slow-motion stall machines such as the Tortoise. Such decoders include the Digitrax DS44, which is an economical design and not intended for twin-coil solenoids.
- Hybrid decoders can provide the appropriate power for both solenoid and slow-motion turnout machines. Examples include the Digitrax DS54 and DS64, together with the ZTC-303.
- Basic turnout decoders are not equipped with additional features such as feedback of turnout position, block occupancy and route setting and cascading.
- There are specialized decoders for specific products such as the DCC Specialities 'Hare', which fits the Tortoise. It has features specific to that type of turnout machine including programmable routing and prevention of turnout run-through. Another example is the Digitrax DS51K1 single decoder for Kato UniTrack turnouts, which have built-in switch machines, or the DS61K1 snap-on for HO scale UniTrack.
- Some come as a combined turnout solenoid and decoder in one package, such as the ZTC-302.

There are numerous differences between the different manufacturers' accessory decoders and some research to obtain the required features is important. Also noteworthy is that many of these decoders require connection to the power bus to receive digital data packets from the command station plus a separate supply to actually power the turnout motor itself. This ensures that all of the power bus current is available for running trains. This is one of the primary reasons why it is a good idea to add an auxiliary 16V ac bus to the layout wiring project, together with its own independent transformer.

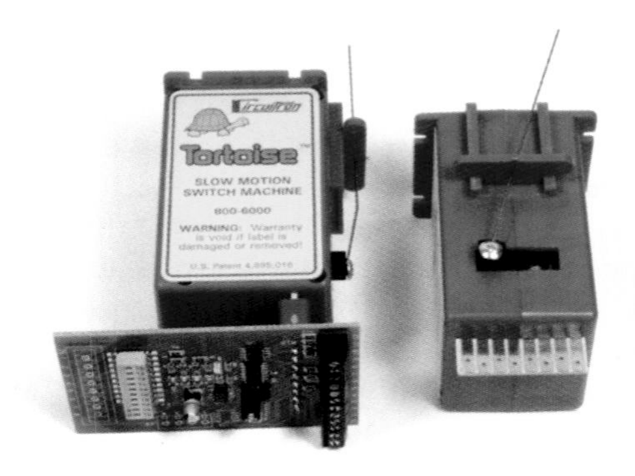


*The Tortoise slow-motion turnout machine is hugely popular owing to its simplicity and reliability. DCC Specialities produces a specialized accessory decoder specifically for this brand of machine called 'The Hare', which fits directly to the motor's circuit board.*

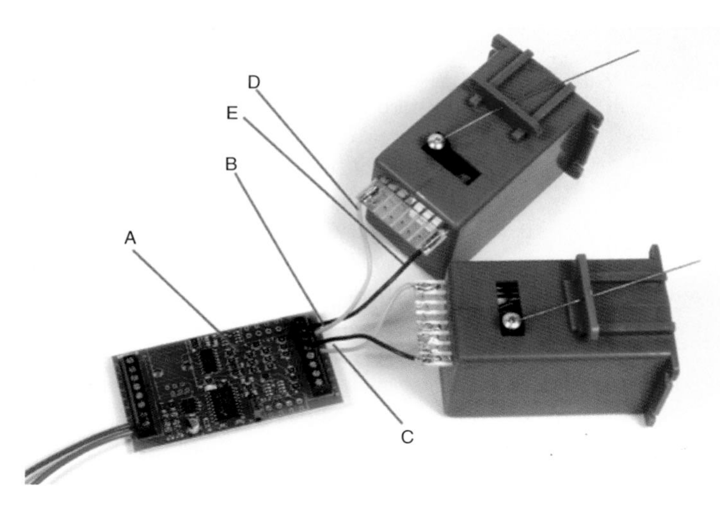


*Also available from DCC Specialities is 'The Wabbit', designed to power two Tortoise slow-motion switch machines. In its simplest guise, the Wabbit (A) is connected to the power bus using terminals 7 and 8 of screw terminal block J1. The connection to both Tortoises is through terminal block J2, terminals 5 and 6 being for one Tortoise (C), with 7 and 8 for the second (B). Terminals 1 (D) and 8 (E) on the Tortoise circuit board must be connected to the correct terminals on the Wabbit. Advanced decoders such as the Hare and Wabbit make possible complex programming for route-setting features, route indication, feedback and 'auto-throw' switching to prevent turnout short circuits from run-throughs.*

When setting up your chosen accessory decoders for turnout control, it is useful to determine which is the 'clear' position of the turnout and which is the 'thrown' or 'diverging' position. The clear route is for the main track or main running line at a junction. The thrown or diverging position describes the turnout as set for the diverging line (logically), although it will not necessarily be the curved route through a turnout. The clear setting will depend on the operation of the layout.

When programming the accessory decoder, look at the position of each turnout relative to the indication on the throttle. Some show a + or – symbol. Others have a small turnout symbol or use the letters C and T for 'clear' and 'thrown'. The + and C symbols indicate that the clear route is set. If the turnout is adverse to that, reprogram the decoder to reverse the position relative to the throttle indication. Some decoders, such as the Lenz LS150 featured in this chapter, have a + and – output for twin-coil solenoids. To reverse the thrown direction of the turnout in question, exchange the + and – wires leading to the turnout. Leave the common return wire alone.

The advantage of doing this check is that, when in operation, seeing which turnout is set for the clear road without having to look at the junction itself is very useful. Also, when programming advanced turnout decoders such as the 'Hare' or 'Wabbit' for route setting, knowing the clear position of all of the turnouts in a route stack is important.

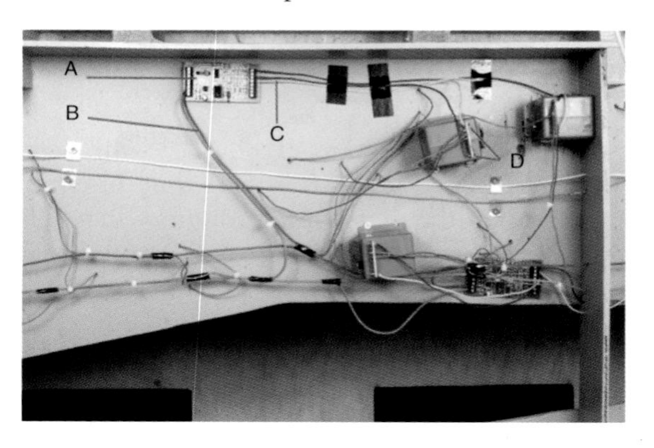


*'The Wabbit' in use on a layout equipped with Tortoise turnout machines. It is fixed to the underside of the baseboard as seen at (A), and is connected to the power bus (B) and the turnout machines (C). Pins 1 and 8 are identified on one Tortoise at (D). Each accessory decoder is programmed with unique addresses for each turnout motor and to create route-setting features.*

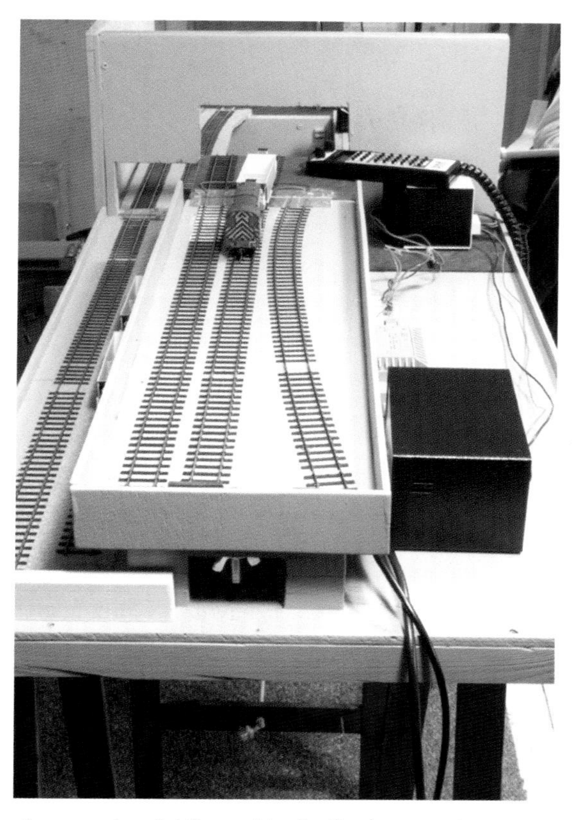

*A sector plate fiddle yard in the final stages of construction for the project layout. The opportunity to provide a clear area for the command station and transformers has also been taken. One of the power bus wires leads up to the sector plate and supplies one rail on each road. The power for the opposite rail will be fed through the latching mechanism, which will keep the sector plate correctly aligned to each running line. When the latching mechanism is withdrawn to allow the sector plate to be moved to align another road up to the running line, the power to that road is disconnected. No switches are used to isolate individual sector plate roads as a result.*

## WIRING THE FIDDLE OR STAGING YARD

Fiddle or staging yards are one area that benefit, from DCC since no switches are absolutely necessary to isolate a train on any road. Sector plate or traverser-type fiddle yards benefit through not having to be divided up into sections either – all that is needed at the very basic level is a connection between the power bus and all the rails. There is a case for providing an isolating switch for one or more roads if there is a need to isolate locomotives equipped with digital sound decoders, so preventing the risk of overheating when the model is 'stabled' for a period of time.

Large staging yards can be wired up to use any inbuilt switch found in some brands of turnout motors (Tortoise) to isolate the current from a given road when the turnouts are thrown against it. This automatic isolation cuts the decoders off from the main supply and can help to protect them when the locomotive is held in the storage road for a period of time. It is particularly useful for locomotives fitted with digital sound decoders and does not require the manual operation of an isolating switch.

## REFINING THE LAYOUT WITH ADVANCED DCC EQUIPMENT

As your knowledge of DCC grows and you become accustomed to operating your layout with a locomotive driving throttle rather than a DC power pack, you will soon come to realize, like me, that there are further refinements you can undertake to improve its operation. You can divide up the layout into power districts using separate boosters that connect to the command station. If the power provided by the base station

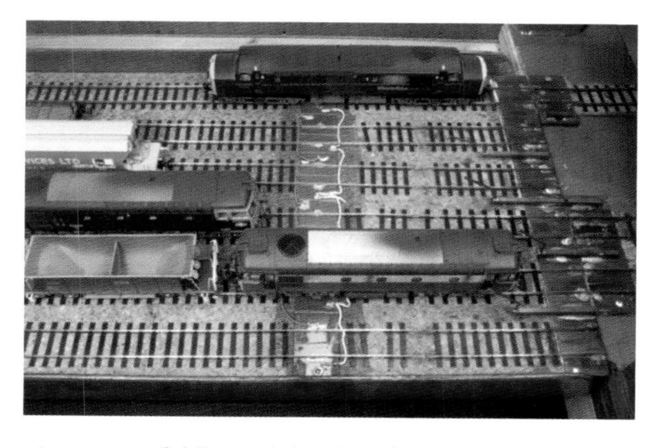

*A traverser fiddle yard showing the connections to each track.*

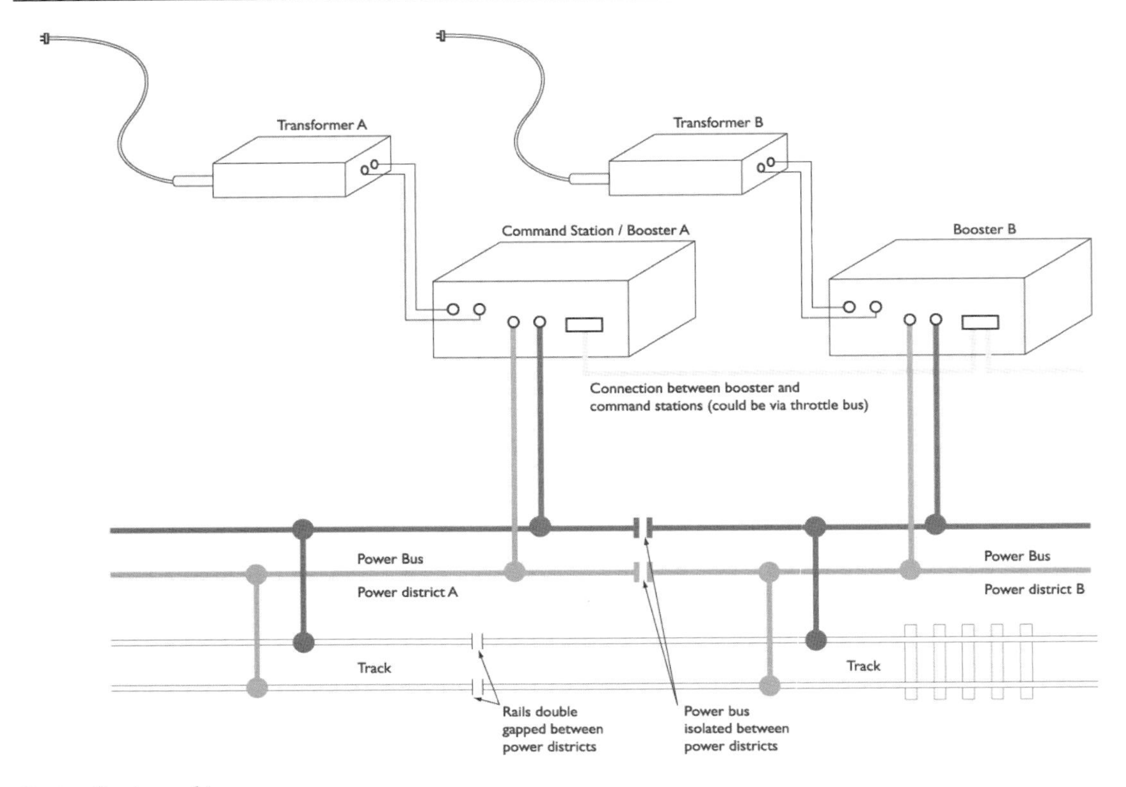


*Power district architecture.*

is sufficient for your needs, you can still improve operations by installing two or more power sub-districts. Auto-reversing modules can be used to power reverse loops, turntables and triangular junctions instead of using a double pole, double throw switch. Then, if you are up to more wiring, consider installing an additional power bus reserved for accessory decoders. This offers many benefits over powering your accessory decoders from the main power bus.

## Power Districts

When you read the specifications for some DCC systems, they appear to feature sufficient capacity to connect vast numbers of throttles to the throttle bus and to operate many locomotives independently. The figures quoted are usually impressively large. Such large figures are tempered by power restraints of the system, unless power districts are built into the design. If the total power output of a DCC system is likely to be exceeded, the layout can be divided up into two or more independent power districts. Each power district is powered by a booster, which in turn will have its own power supply. The boosters are connected to the command station through the throttle bus network or other method, depending on the system, so that the command station retains control of the whole layout and permitting use of the large operating capacities of some systems.

Some modellers consider that ideally you should create power districts for distinct working areas of the layout, such as a large classification yard, the main running lines and perhaps the fiddle yard, too. It depends on the objectives and size of the layout, together with the necessary power required to keep the layout operational. They are sometimes used to prevent a short circuit due to a turnout run-through or derailment from shutting down the rest of the layout, disrupting the enjoyment for other operators.

Power districts are completely independent of each other and should be equipped with independent power bus wires. The tracks crossing from one district to the next must be double gapped at the boundary maintaining the independence of each district. The rails have to be in sync with each other too, which is why they are usually given an identification: rail A and rail B, for example.

Let's say a layout is to be powered by three power districts. The first thing to note when making up a shopping list of equipment is that the command station base station and its booster will form the first power district. Only two additional boosters and transformers will be required for the second and third districts.

## Power Sub-Districts

When you first invite your model railway friends over to admire your new DCC layout, and perhaps start the first of many enjoyable operating sessions, one thing about working with multiple operators will quickly become apparent. As you speed an express freight along the mainline, concentrating on the track geometry and controlling your train, a derailment in a yard operated by someone else causes a short circuit and suddenly shuts the command station down. Everything grinds to a sudden halt. Sound decoders go silent and lights are extinguished. The concentration required for multiple train running is destroyed by the delay as the offending stock is re-railed, or a locomotive has to be pushed back from a turnout that was set against it before the reset button can be pressed, restoring everything back to where it was!

The problem is exacerbated if the turnouts and points on the layout are connected to the power bus and so the misaligned switch cannot be properly set until the short has cleared and the system has rebooted. How can you avoid being affected by the carelessness of your yard shunter?

An interesting development is the use of specialized circuit breakers installed on the power bus to divide the layout up into 'power sub-districts'. They work on the same principle as power districts in that they are used to create independent electrical blocks in the power bus and can be used to separate operating areas of the layout

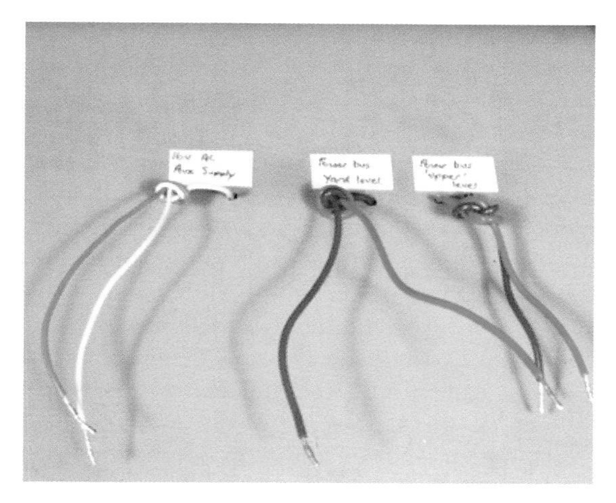

*Power bus wires (green and red) emerge from under the layout at the point where they will be connected to the base station. They are clearly labelled for identification. The upper and yard areas of the project layout are wired independently of each other to create two power sub-districts.*

such as classification yards, the main running lines and station areas. Boosters are not required to create power sub-districts and the power is limited to that supplied from the DCC base station.

Circuit breakers such as the Power Shield series are used to detect a short circuit and cut the power in a given sub-district without affecting the others. This means that a short in the fiddle yard won't affect the classification yard. They also provide advanced short-circuit protection for the DCC base station and any

*This four-circuit breaker from Digitrax, which includes a LocoNet connection, can be used to create four separate power sub-districts.*

boosters in use on the layout. Power districts powered by a booster can also be subdivided into power sub-districts if required.

The same rules apply: for the circuit breakers to be effective, each power sub-district must be independent of the other with rails double-gapped between sub-districts. On the project layout, the upper station area and the lower yard level formed two power sub-districts using a two-breaker PS2 Power Shield device supplied by Tony's Trains Exchange. Ensure that the circuit breakers can handle the power spikes during start-up of the system, especially when power-greedy devices such as sound decoders are in use.

### Accessory Power Bus

An extension of the power district and power sub-district wiring method worth considering, and recommended on large layouts, is an independent accessory power bus with its own power supply and booster. This would be a second set of power bus wires running around the layout, which is connected to all the accessory decoders that would otherwise draw both current and DCC signals from the main power bus. There are several advantages in this additional wiring:

- A short circuit on one part of the layout (operator error again?) will not affect the accessory decoders and turnouts can still be operated to clear a run-through and reset the system.
- The power required for power-hungry accessories such as twin-coil solenoid point machines can be provided by a separate booster and power bus, maximizing power in the power bus supplying the track and locomotives. In extreme cases, twin solenoids will drain all of the power from a power bus when more than one is fired for route setting and may even look for additional voltage from the throttle bus when competing for power with locomotives. Separating the track supply power bus and that supplying accessory decoders is regarded as good practice.
- Creating a separate power sub-district for an accessory bus on small layouts with low power demand will still provide a degree of protection from shorts, enabling turnouts to be changed even when a locomotive runs a turnout, thus resetting the system when the short circuit clears.

It is worth noting that some accessory decoders, such as the Lenz LS150, require their own power supply, only using a connection with the power bus to receive DCC signals. For that a separate ac power supply should be run along the length of the layout, with its own transformer.

### Auto-Reversing Modules

These devices are designed to detect a reversal of the track supply and will automatically reverse the polarity before the command station shuts down. They are wired into reverse loops or any situation where a reversal of track power polarity is likely and may be regarded as automatic, 'install-and-forget' devices. Many power protection devices, such as the Power Shield circuit breaker devices, are also automatic reversing loop devices and can be used as such.

## MORE ON LAYOUT WIRING

### Programming Track

Plan for a separate programming track in the fiddle yard or somewhere convenient on the layout. Programming tracks allow the programming of locomotives by the command station, provided through dedicated output ports, and will detect errors in decoder installation together with faults with the decoder itself. The output power from some systems is not so high as the main power, preventing further damage to the decoder. Some modellers use a double pole, double throw (DPDT) switch to switch a siding or spur track from main power to programming power, enabling the operator to drive a locomotive into the siding, re-program it and then drive it away again.

It is worth noting that many entry-level systems only offer 'programming on the main' by not providing separate outputs for an independent programming track. Programming using this method with some entry-level sets is

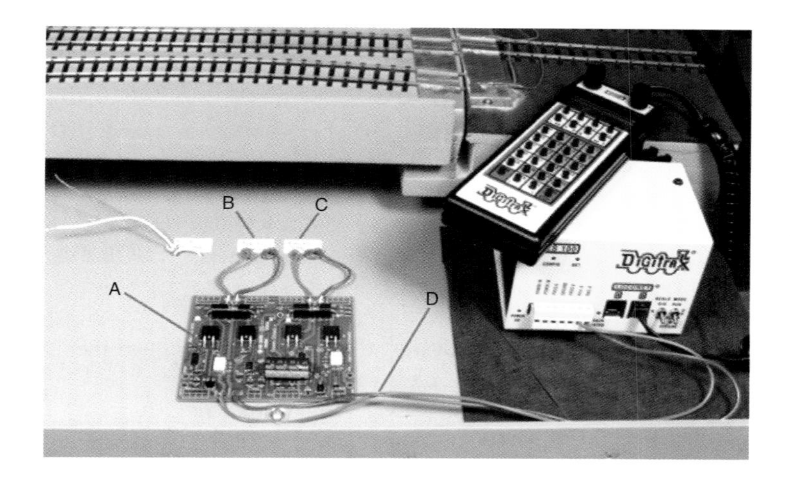


*Connecting a circuit breaker. This is the PS2 Power Shield type from Tony's Trains Exchange, which has been used on the layout to connect the upper and yard power sub-districts to the Digitrax base station.*

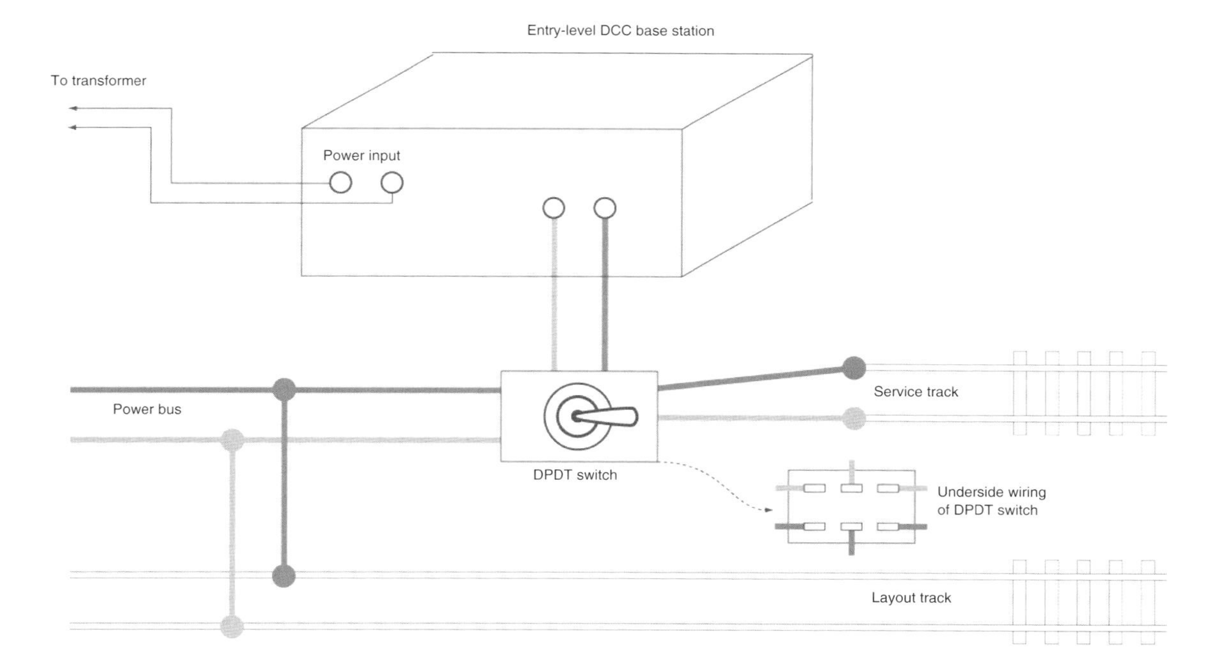


*Setting up a service track for entry-level systems that broadcast programming commands.*

not targeted to one locomotive (loco specific), identified by entering its address, but broadcast through the whole layout. The programming instructions will be taken up by all of the locomotives present on the layout.

When a loco has to be programmed in this way, the system may have to be disconnected from the layout and connected to an independent piece of track or, alternatively, by removing all locomotives from the layout so the broadcast programming is only received by the target locomotive. Having a dedicated siding or programming track with a DPDT switch to redirect the power bus power to an isolated siding or a programming track is the best solution, avoiding the need to remove other locomotives from the layout.

Programming on the main using advanced sets is more specific, targeting the one model whose address has been entered in the system. There will be more on this later (*see* Chapter 4), but an independent track for such programming is not required.

## OUT WITH THE OLD, IN WITH THE NEW: ADAPTING EXISTING LAYOUTS

An analogue layout wired for cab control can quickly be adapted to run with a DCC system, provided the wiring has been completed using safe wiring practices. If poor-quality wire has been used to supply current to the track, then a rewire should be considered before embarking on a switch to DCC.

The transformers and controllers used on the original control system should be removed (except those used to control turnouts and signals) and the DCC base station connected to the wires that connected the original dc controllers to the electrical block switches. All of the switches should be turned on to ensure that all of the track is live. A problem you may encounter is the need to provide additional track feed wires. Eventually you may also have to make reliability modifications to Peco Streamline turnouts. This should always be considered as a temporary measure and a complete rewire of the layout undertaken at the earliest opportunity.

I personally have my doubts about mixing DCC with analogue control on the same layout by using some sort of change-over track section between the two system's districts. This is prone to error that could damage the DCC base station and other devices too. I personally do not think it is safe, since it goes against all the principles of safe electrical practice and should be avoided, even for large layouts. Decide if DCC is for you, then commit to it wholeheartedly. Don't take a halfway house approach – not even temporarily!

## ELECTRICAL FAULTS AND FINDING THEM

At the beginning of this chapter I stated that wiring a layout is simply that and there is no such thing as 'wiring for DCC'. It's an often repeated mantra by the DCC manufacturers too, here stated by The Model Rectifier Corporation's DCC literature:

> There are plenty of books available through the hobby press on layout wiring. Whether the book is recent and is about layout wiring for DCC, or if the book is 50 years old and deals with cab control and block sections, proper wiring is proper wiring. That's it! More problems can be traced to poor wiring practices than anything else.

That is an important statement: the basics of layout wiring described in most books will cover you for your DCC layout when read in conjunction with the manuals. After all, you can hook up a DCC system to a dc cab control layout and have a decent chance of seeing it all work first time provided good wiring practices have been followed. All manufacturers stress the importance of good wiring practice. Digitrax states in its 'Super Chief' DCC manual: 'Remember, no matter how you run your trains, you should always use safe wiring practices.' That manual is one of the best on the market and a great resource for any DCC user.

No matter how careful you are, there is always the chance of a fault occurring, especially when you plug the system into the layout for the first time. Whilst DCC makes dividing your layout into small electrical block sections unnecessary, some modellers prefer to divide the layout and therefore the power bus into large block sections to assist with fault detection. This can be done with switches – they could be left under the layout given that they will not be used regularly (or, fingers crossed, not at all!) rather than mounted on the layout fascia where they could be accidentally switched, or installed on a control panel with the associated long runs of wire, which is definitely not desirable and defeats one of the benefits of DCC. Incorporating circuit breakers will also help with fault detection.

Portable layouts are naturally divided into blocks – each baseboard is a block, electrically speaking. The electrical connections between boards are natural isolators. Each one is connected to the next with jumper cables and connector assemblies, which are unplugged when the layout is dismantled for transport or storage. If a fault occurs during operation, the connectors can be unplugged to help isolate the fault to one of the boards.

One of the most common faults on DCC layouts is a short circuit that shuts down the command station. There are obvious causes that are also temporary in nature, such as out of gauge rolling stock wheels bridging the gap between stock rails and switch rails on some types of turnout. Another one usually associated with new layouts at the construction stage, utilizing hand-built track or PCB copper-clad sleepers, is the incomplete isolation gap in a single sleeper in a turnout, somewhere on the layout. This has to be laboriously tracked down and the offending gap properly filed through the copper veneer. Another that has caught me out is the tiny fleck of copper or burr of nickel silver rail bridging an insulation gap – again vigilance soon sees such faults removed.

Turnout machines can cause short circuits when they do not throw correctly, resulting in the built-in polarity change switch bridging both terminals. The short takes place through the internal motor switch and is not easily detected. Shorts occurring when a locomotive runs over a turnout crossing vee may mean that the polarity change switch is wired up the wrong way round, or the crossing vee on live frog turnouts is not isolated from the diverging end of the turnout.

It's always worth doing preventative checks under the layout to see if exposed wire can come into contact with another, where direct-soldered connections to the power bus have been made. The wires under portable layouts are subject to more than normal stress when they are moved and that may be enough to shift insulation and cause unwanted electrical contact. Permanent

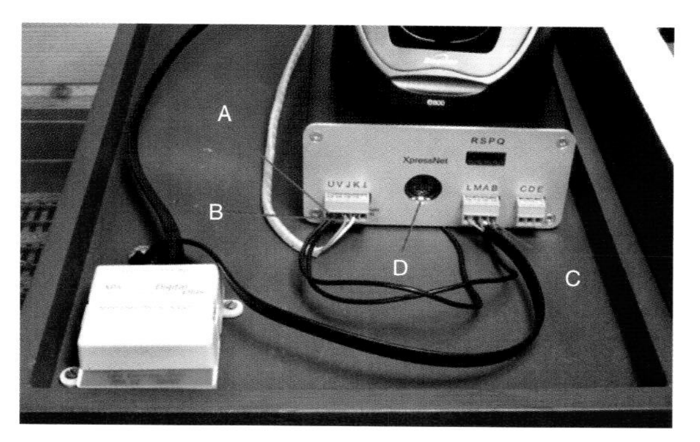


*Making the final connections: a power supply (A) from the transformer, the power bus connection to the layout (B), the throttle bus connection (C) and sometimes the throttle port in the back of the base station comes in handy, too (D).*

layouts in basements, lofts and outhouses have space underneath the baseboards that finds favour for storage – all those boxes and stacks of magazines. Retrieving a box or pushing one back can potentially disturb wiring.

One of the simplest faults is when a track feed wire becomes disconnected from the running rail it was soldered to and that rail suddenly becomes dead. The first sign of trouble is when your trains stall on that rail section. Sometimes the disruption in power is not easily identified as the wire may appear to be in contact. A dry solder joint is all it takes. At least you do not have to isolate blocks of layout to locate the fault when this happens. Care with soldering technique will help minimize such occurrences.

## The Coin Test

A simple but vital test to see if you have good power supplies throughout the layout is to use a coin (or other metal object) to induce a short circuit by placing it across the rails. If the power supply is good, the command station/booster will trip out in a flash (not literally we would hope!). Failure of the short-circuit detection to trip power out means that the power supply is not as good as it should be and this could be down to insufficient feeders between the track and the power bus.

CHAPTER 4

# Mobile Decoders: How to Fit and Test Decoders in Locomotives and Multiple Units

*Decoders come in all shapes and sizes, with differing harnesses and number of functions. After all, there is no 'one-size-fits-all' locomotives decoder – a Hornby 'Britannia' and a Class 60 may be similar in size, but the former is relatively short on interior space and may need a different solution to the Class 60.*

## INTRODUCTION

Mobile locomotive decoders (sometimes called controllers or receivers) are ingenious devices that read and act upon the data packets transmitted by the command station to control our model locomotives. In effect, decoders are control computers small enough to fit the tight confines of our models (N gauge modellers will be all too aware of how little space there can be), yet powerful enough to cope with track power and to provide power for the motor.

Decoders are capable of doing a multitude of tasks. They interpret the data packets transmitted

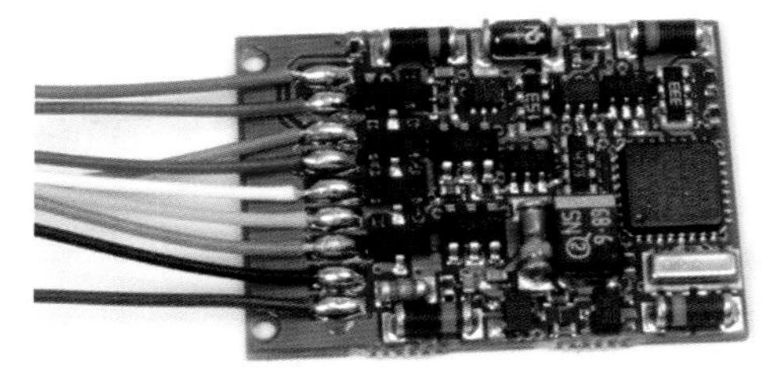

*A typical single-sided decoder with soldered wiring harness. This is a four-function decoder with nine harness wires. Wire colours are (generally) standard for all decoders.*

by the command station (as directed by the throttle) and will act upon the commands if their unique address is attached to the data packet. A locomotive (mobile) decoder will rectify the ac waveform current to 12V dc so it can be used by the motor and onboard systems such as lighting effects, digital sound and steam generators. The decoder acts as a controller, switching onboard systems on or off and controlling the current to the motor for movement and changes of direction if it is commanded to do so by the command station. It will do these things independently of any other decoder on the layout unless commanded to work with one or more other decoders in a 'consist'. The decoder can be 'programmed' to perform particular functions too, such as different lighting effects: strobes, flashing or dimming. Decoders will perform braking delay and inertial effects together with a variety of other operational features.

When you consider the size of a typical OO/HO scale decoder, that level of performance is quite remarkable. Yet, when you examine one, its appearance is unremarkable, just a printed circuit board (usually greenish in colour) with tiny components dotted all over it – rarely anything more than that.

Decoders come in all shapes and sizes and not all decoders are the same for a given scale:

- Double-sided decoders have components on both sides of the circuit board. They may be smaller in area than a single-sided decoder but thicker, making them useful for some tight spots.
- Single-sided decoders have components on one side only, the opposite side being flat. They tend to be of greater area but thinner, typically less than 3mm in thickness.

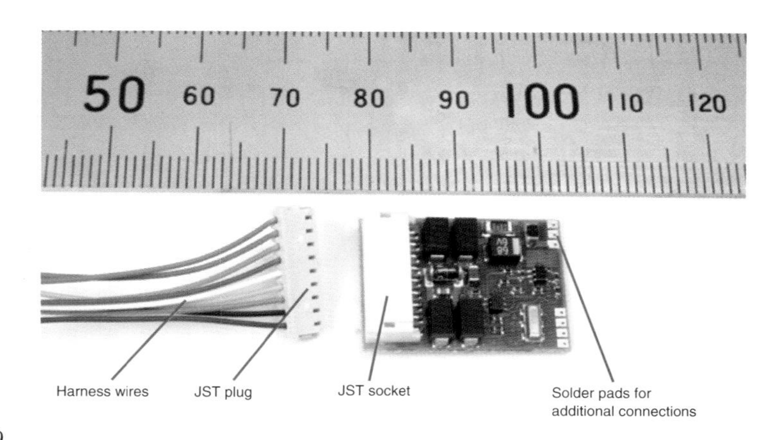


*A JST plug and socket fitted to a double-sided four-function HO/OO decoder.*

- Some decoders are bespoke and are designed to fit specific models without soldering in most cases. Bespoke decoders are particularly popular in US-outline modelling where the lighting circuit board is completely removed and replaced with a bespoke decoder that is exactly the same size and shape, even including LEDs for running lights. While some decoders are available for specific HO scale models, this practice is much more common in N gauge (N scale) where the space occupied by the printed circuit board is ideal for accommodating the decoder in a model that otherwise will have little or no space.
- The manner in which the wiring harness is attached can also vary between decoders. Some have soldered connections in which the wires are soldered to tiny solder pads along one edge of the decoder. The impact on installation space is usually slight against having to de-solder the wiring harness from the various points in the model if it is to be removed for any reason. Some decoders are factory fitted with JST sockets soldered to one edge of the decoder. The corresponding harness has a plug that slots neatly into the JST socket. This makes it easy to unplug the decoder from the harness wires should that be necessary. That benefit should be set against the disadvantage that JST plug and socket arrangements take up more space than soldered connections.
- NEM 651 and NEM 652 (NMRA RP-9.1.1) DCC interface plugs may be fitted to the opposite end of the wiring harness, making it easy to plug the decoder straight into a DCC-ready locomotive (one equipped with a socket

*Direct-fit decoders without a wire harness are growing in popularity because they can be plugged directly into a NEM 652 socket without having to find room for harness wires. This picture shows a ZTC4007 fitted to a Hornby 'Royal Scot' locomotive.*

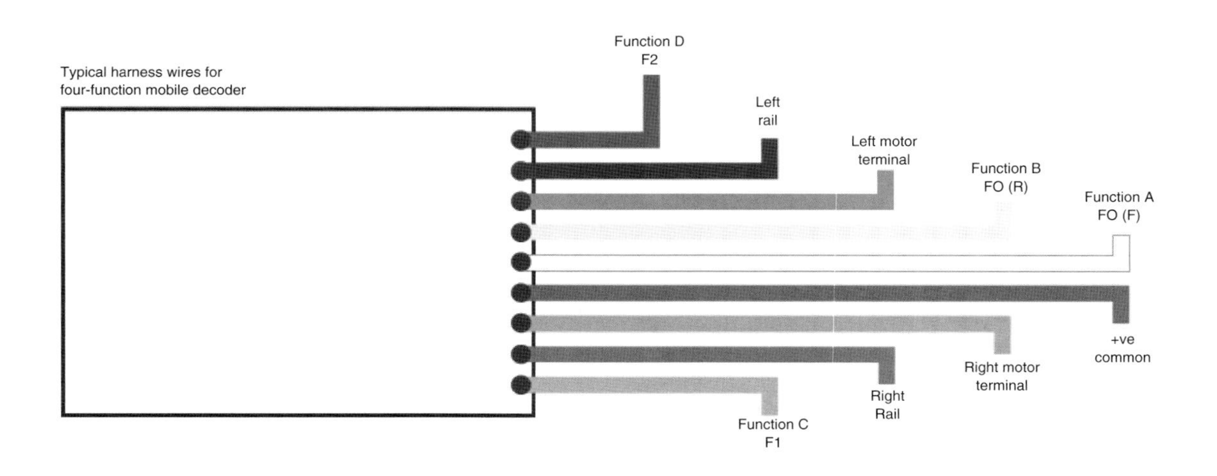


*Decoder harness wire identification.*

to NEM standards or those described in NMRA RP-9.1.1).

- Direct-fit decoders without any harness are growing in popularity. They are of benefit when space is tight, sometimes too tight for even a short wiring harness. A NEM plug is connected directly to the decoder board. Three basic types are available: NEM 652 eight-pin, 21-pin and NEM 651 six-pin decoders.
- Large-scale decoders for O gauge have greater power ratings and (normally) no wiring harnesses. A set of screw connectors is fitted to one edge of the decoder.
- The number of harness wires (or connections) is dependent on the number of functions and not always the same for every decoder. However, the wire colour codes are standard, as specified in NMRA recommended practice RP-9.1.1.
- Ratings vary between decoders and there are two figures of which you should be aware: the continuous rating in amps tells you the maximum constant current draw that the decoder can deliver, while the 'peak' rating is the maximum current draw that the decoder can withstand for short periods of time, notably the motor current 'spikes' when starting a heavy train. The current draw of the motor must not exceed that of the decoder. (*See* table of typical current consumption for ready-to-run locomotives, page 79.)
- Function decoders only control onboard systems such as lighting and sound. They do not have any motor control output. The typical mobile function-only decoder will have four or six functions but no orange or grey harness wires for connection to motor terminals. They are used in trailer cars in multiple units, unpowered 'slave' locomotives or as additional controllers to the mobile locomotive decoder in locomotives with sophisticated running lights, such as ditch lights or strobes, in addition to the normal running lights. In that application, the functions will need to be 'remapped' to higher function buttons on the throttle to avoid conflict with F0(F), F0(R), F1 and F2 buttons used for controlling the output of the main decoder.

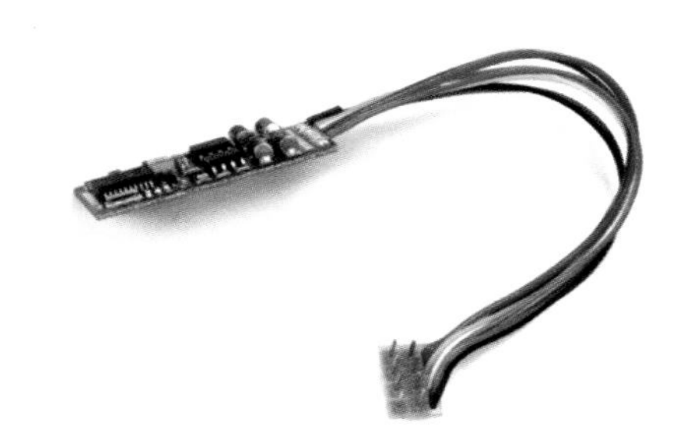

*A typical single-sided decoder with 8-pin plug attached to the wire harness.*

## POWER RATING OF DECODERS

Decoders have power ratings, just like any other electrical component. Those power ratings are often quoted as a continuous rating and a peak rating, the latter reflecting its ability to protect the decoder from momentary current spikes when a locomotive starts with a heavy train.

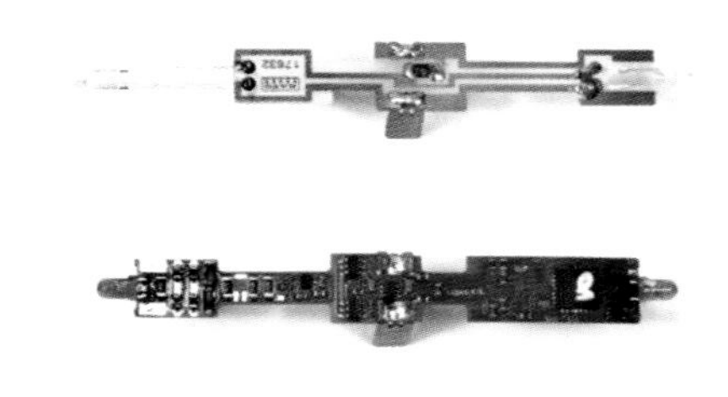

*Some decoders are bespoke, designed to fit a specific model. US N scale modellers enjoy the benefits of bespoke decoders that have the same shape as the dc lighting circuit board supplied with the model. The one shown in the bottom of this photograph is a Digitrax DN163K1B designed to fit in US N scale locomotives such as Kato SD70, AC4400CW and C44-9 models. The original circuit board is shown for comparison.*

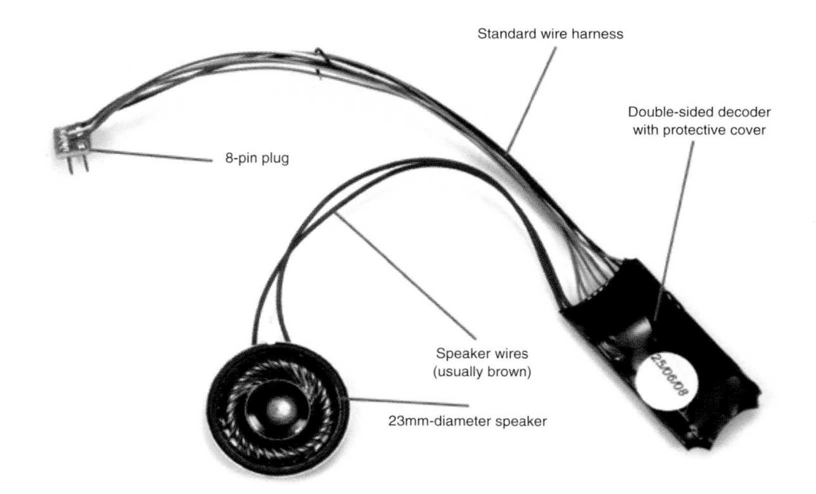


*Decoders can be very specialized, such as this HO/OO digital sound decoder produced by ESU. Not only does it contain the functions found in 'regular' four-function decoders, but it has sound file storage and an amplifier too!*

The total power rating of some decoders usually includes the power rating of each function output (usually 100mA each) in the total, so that has to be allowed for when choosing decoders for a model. Matching decoders to specific models is not always as straightforward as selecting one of the correct size. While the usual rating of a decoder for HO/OO gauge may be 1 amp, well within the average of 0.4 amps for a modern OO gauge engine, some models will draw considerably more than that and Heljan uses motors that can easily exceed that during free-running. You also have to consider what might happen if the model stalls or slips to a stand on a heavy train. The current consumption can double in no time.

It is worth rigging up a simple test meter assembly to test the current draw of a model when free running, say on a rolling road. This will help you to determine if the model draws excessive current due to a power-hungry motor or an electrical fault. Either way, you may have saved a decoder from burnout. It works the other way too – some small tank locomotives, shunters and switchers equipped with small motors and simple mechanisms may need only a 0.5 amp micro decoder and usually that helps, since such locos have little spare room for additional components.

## DCC-READY – WHAT DOES THAT MEAN?

The term 'DCC-Ready' sometimes causes confusion among novice modellers because it suggests that a model is ready to go on a DCC layout. When manufacturers apply that term to their models, they are indicating that they are equipped with NEM 651 or NEM 652 (NMRA RP-9.1.1) sockets that will accept a decoder fitted with the corresponding plug. They are plug and play, but a decoder has to be obtained and fitted.

Models factory-fitted with decoders are usually labelled 'decoder onboard' or the box text will state the type of decoder installed at the factory. Usually, DCC-onboard models are supplied with a brief instruction manual and a default decoder address of 0003.

## INSTALLATION ESSENTIALS

Decoder installation is basically simple to do. The principle is to connect the decoder so that it sits on the electrical circuit between track power (current pickups) and the motor. When the decoder is in place there should be no direct connection between track power and the motor to avoid damaging the decoder – that is

## POWER CONSUMPTION OF SELECTED PRODUCTION MODELS

| Locomotive | Manufacturer or label | Scale/ Gauge | Date of test | Power consumption amps | |
|---|---|---|---|---|---|
| | | | | Free running | Slipping |
| Class 04 shunter | Graham Farish | 1:148/N | July 2007 | 0.04 | 0.05 |
| Class 37/4 (2007) | Bachmann | 1:76/OO | July 2007 | 0.21 | 0.44 |
| Class 37/0, 37/4 | ViTrains | 1:76/OO | March 2007 | 0.28 | 0.85 |
| Class 57 | Graham Farish | 1:148/N | March 2007 | 0.19 | 0.24 |
| Class 170 'Turbostar' | Graham Farish | 1:148/N | January 2007 | 0.18 | 0.21 |
| N2 steam locomotive | Bachmann | 1:76/OO | January 2007 | 0.12 | 0.19 |
| Class 08 shunter | Hornby | 1:76/OO | November 2006 | 0.20 | 0.24 |
| GP38-2 | Athearn | 1:87/HO | August 2006 | 0.36 | 0.47 |
| GP38-2 | Proto 2000 | 1:87/HO | August 2006 | 0.18 | 0.35 |
| Class 67 | Hornby | 1:76/OO | October 2006 | 0.18 | 0.28 |
| M7 steam locomotive | Hornby | 1:76/OO | September 2006 | 0.16 | 0.22 |
| 4MT steam locomotive | Bachmann | 1:76/OO | September 2006 | 0.10 | 0.19 |
| Class 50 | Hornby | 1:76/OO | August 2006 | 0.36 | 0.66 |
| Class 66 | Graham Farish | 1:148/N | June 2006 | 0.20 | 0.29 |
| Class 60 | Hornby | 1:76/OO | August 2006 | 0.33 | 0.69 |
| 45xx steam locomotive | Dapol | 1:148/N | August 2006 | 0.11 | 0.16 |
| Class 73 | Dapol | 1:148/N | February 2006 | 0.19 | 0.28 |
| Class 66 | Dapol | 1:148/N | April 2006 | 0.38 | 0.41 |
| Class 44 | Graham Farish | 1:148/N | March 2006 | 0.20 | 0.30 |
| Class 43 HST power car | Hornby | 1:76/OO | February 2006 | 0.14 | 0.33 |
| 57XX Pannier Tank | Bachmann | 1:76/OO | March 2006 | 0.10 | 0.15 |
| Class 91 | Hornby | 1:76/OO | March 2006 | 0.19 | 0.30 |
| Class 33 | Heljan | 1:76/OO | February 2006 | 0.42 | 0.76 |
| Class 04 shunter | Bachmann | 1:76/OO | March 2006 | 0.14 | 0.20 |

absolutely mandatory. A safe space within the confines of the locomotive is chosen upon which to fix the decoder using a spot of glue or a double-sided sticky pad. It is important that the decoder does not come into contact with any metal surfaces and it is considered good practice to cover adjacent metal surfaces with insulation tape if there is any chance of accidental contact.

Each decoder has colour-coded harness wires, the colour being the same for every NMRA compliant decoder and relating to a specific connection in the model – not all of them are necessarily used. Taking a typical four-function OO/HO gauge decoder, the harness should consist of nine coloured wires. The red and black wires are connected to track power supply via the current collection devices. The orange and grey wires are connected to the motor terminals. That's it for a basic installation.

The remaining five wires relate to the control of onboard systems such as lighting. The blue wire is a positive common power supply that can be connected to the positive side of any lighting system. The four functions are controlled by the remaining four wires, which are connected to the negative side of any lighting system.

### Points to Remember

- Never wrap a decoder with insulation tape or a heat shrink sleeve to protect it from

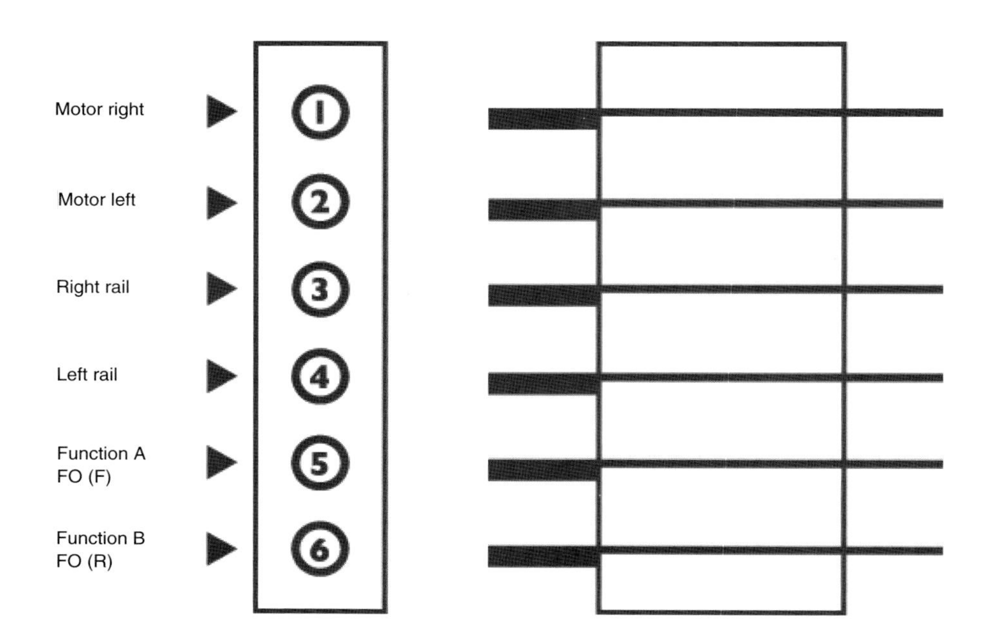


*NEM 651 DCC 6-pin interface socket arrangement.*

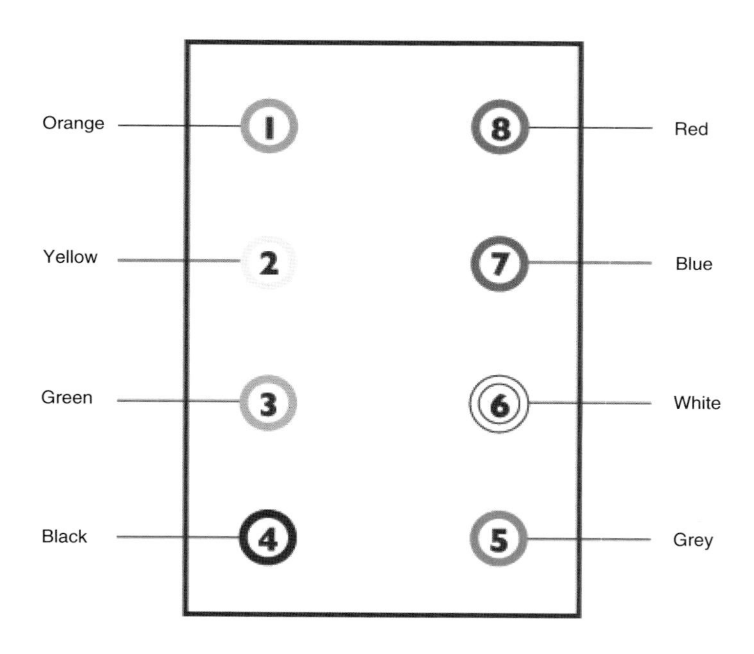


*NEM 652 DCC 8-pin interface socket arrangement.*

accidental contact with metal. Decoders need a flow of air over them to function properly and insulation tape will prevent that from happening. It is better to apply insulation tape to adjacent metal surfaces instead. Some decoders are supplied with factory-fitted plastic sleeves. Although these will provide some protection from metal, it is more likely that the sleeves are present to prevent static electricity damage. Those decoders supplied without sleeves usually do not need protection from static electricity. However, the often quoted reason for a decoder not being fitted with a protective sleeve is that the decoder usually runs hot and the sleeve would cause overheating. This is another urban myth and is rarely the case.

- The installation of decoders is often complicated by factors such as split frame chassis, as found in most N gauge locomotives, or a simple lack of space for the accommodation of a decoder without risk of damage when a loco body is refitted. Lighting systems will also make a decoder installation project longer to do. We will deal with these issues further in this chapter. (For matters relating to lighting, *see* Chapter 5.)
- An important factor that modellers should consider is that a high specification decoder, with features such as 'silent drive' or 'back EMF', will not improve the performance of poorly maintained or ailing models. The model should be in perfect working order with clean wheels and current pickups before the decoder is fitted. In the case of new models, they should be fully run-in, cleaned and freshly properly lubricated before undertaking a decoder installation and placing them in service. This is how the model can be prepared for service. Consider it to be part of the game: the full-size railways commission their new locomotives and stock, so why shouldn't you?

## RUNNING-IN AND INSPECTION

As exciting as buying a new model can be, you should resist the temptation to plonk it straight on your layout and run it as you would an established locomotive in your collection. All mechanisms benefit from a couple of hours of running under controlled conditions to break them in gently and to check that there are no defects that need attention or a return to the manufacturer for a warranty repair. Additional lubricant may be required during the running-in process because you don't know how long that model has sat on the shelf, and factory-applied lubricant can dry out over time, such as when the model was in transit or stored in a model shop before being sold to you.

The first stage is to read the service leaflet before looking the model over to see if there are any loose parts or components that could cause damage if they fell off during operation. Sometimes it may be necessary to return the model to the model shop should you discover damaged detailing parts or something visually wrong with the mechanism. Bogie sideframes and base plates can be removed so that the gears may be checked for fluff, hairs, dirt or particles of metal that will cause wear and indifferent running. Also check that there is sufficient lubrication on the bearings before you run the model. If poor running persists, examine plastic gears for burrs and moulding pips.

The visual inspection continues by checking the wheel sets to see that they comply with accepted standards for gauge and back-to-back measurements. This examination is undertaken

*Running-in and careful testing of new acquisitions to the loco roster is important. You can avoid many post-installation problems this way. A Bachrus rolling road is used to test a Proto 2000 HO scale switcher.*

with a back-to-back gauge such as that specified by the scale societies or the NMRA and adjustments made if necessary.

Once satisfied that the model is correctly gauged and in good order, power is applied to the model via leads from a simple analogue controller. This is to check that the wheels are concentric and there are no alarming squealing, grinding or grating noises coming from the mechanism. Such symptoms should be investigated and if they cannot be identified, the model must be returned for repair.

It is at this stage that running-in can commence and a rolling road for static running and testing is ideal. The model is placed on the saddles and run-in at half speed for fifteen minutes in either direction while under supervision. If there are no signs of distress, the model is run for a further fifteen minutes in either direction at near full power. Usually, with the quality of assembly of modern models, all is well and defects are rarely found.

I never leave my models unattended when running on a rolling road because, should the mechanism bind, I need to be present to take immediate action to prevent the motor from burning out or the mechanism from becoming damaged. My service track is placed on a shelf above my workbench and running-in is undertaken when I am doing other modelling tasks, so the process is supervised at all times.

The model is returned to the inspection cradle and the bogie frames or base plates removed again to check the level of lubrication to see if it has worked into all moving parts and bearings. Remove the body and inspect the bogie gear towers and drive shafts too. Dirty grease and oil is cleaned away because that may contain particles of material that will cause excess wear in the future. Excess lubrication should be removed and those areas that appear to lack lubrication should be gently oiled with a tiny drop of model mechanism oil. Reassemble the model and subject it to a further fifteen-minute period of running-in, which can be extended if required. Some take longer to settle down than others, but you will hear the change in the speed of the model for a given voltage as running-in progresses. A smoothly running model will benefit from all the clever electronics in a decoder and you will enjoy operating it all the more.

## TOOLS AND MATERIALS

This section deals with the fitting of decoders to 'easy to fit' models and to locomotives with known practical issues to demonstrate the principles of decoder installation and some ideas on how to work around practical difficulties. You should have a number of tools and materials to hand on your workbench. Indeed, it is useful to assemble a decoder installation toolkit, perhaps a plastic food container with the essential supplies and tools dedicated to the task. The disadvantage of duplicating your tools is more than offset by being able to put your hands on the kit that you need straightaway, thus saving time in searching for that elusive roll of insulation tape.

- Soldering iron and cored electrical solder.
- Solder pump or braid for the removal of excess solder.
- Heat shrink installation sleeve of 1.5mm and 2.5mm diameter.
- Insulation tape.
- Wire strippers.
- Pliers.
- Tweezers.
- Self-adhesive sticky pads for securing decoders.

*Useful tools and supplies to have to hand when fitting decoders.*

- Modelling knife and spare blades.
- Jeweller's screwdriver set for releasing body securing screws.
- Scraps of 20thou styrene card for insertion between the locomotive body and chassis to hold the body away from chassis retaining clips.

## SIMPLE INSTALLATIONS

The simplest installation you can do is to fit a decoder to a locomotive fitted with a DCC interface socket. In a OO or HO scale model, that is usually the eight-pin NEM 652 socket or a six-pin NEM 651 socket in N gauge models. After the model has been tested, simply remove the bodyshell to gain access to the interior. Locate the socket and remove the dummy plug, which should be stored in a safe place should it be needed again. The pins are numbered and correspond with specific leads on the decoder harness. Socket 1 is usually identified on the circuit board and the decoder plug should be oriented so that pin 1 (orange wire) is plugged into the socket hole marked No.1. Simple!

Locate a safe place to accommodate the decoder where it makes no contact with exposed metal surfaces and stick in place with a double-sided sticky pad. The model should be placed on the service track and tested to see that the installation is safe. This is done by checking the decoder address (CV1), which should read back on the throttle display as the factory default setting of 0003. This means the installation is safe and a new address can be chosen and entered into CV1. If the installation is incorrectly done, the readout will show an error code. The operating manuals relating to your chosen system should be read to find the meaning of the error code and how to take appropriate action to correct the error.

Even though this seems like a very simple way of installing decoders, there are some pitfalls. Plug the decoder in the wrong way round and the model will run 'backwards'. The lighting systems will malfunction too. The cure is simple – plug the decoder in the other way round! Don't worry if you do this – the decoder is unlikely to fry and, believe me, all experienced DCC users have done this at some time when fitting decoders in a hurry! A more serious problem can arise when the soldered points under the circuit board make accidental contact with the metal chassis frame. This is a particular problem with Heljan models, which have little clearance between the underside of the circuit board and the chassis. Remove the circuit board and apply insulation tape to the solder points as

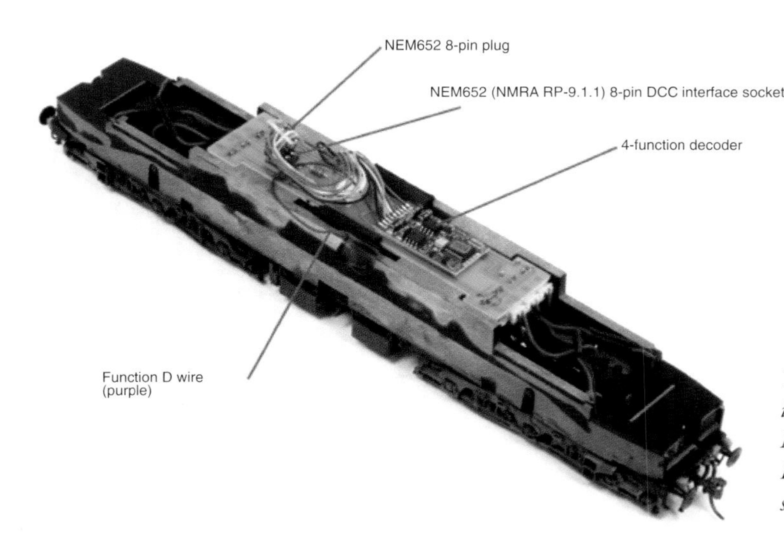


*The simplest decoder installation involves models fitted with NEM 652 interface sockets. Remove the dummy plug and simply plug your decoder in.*

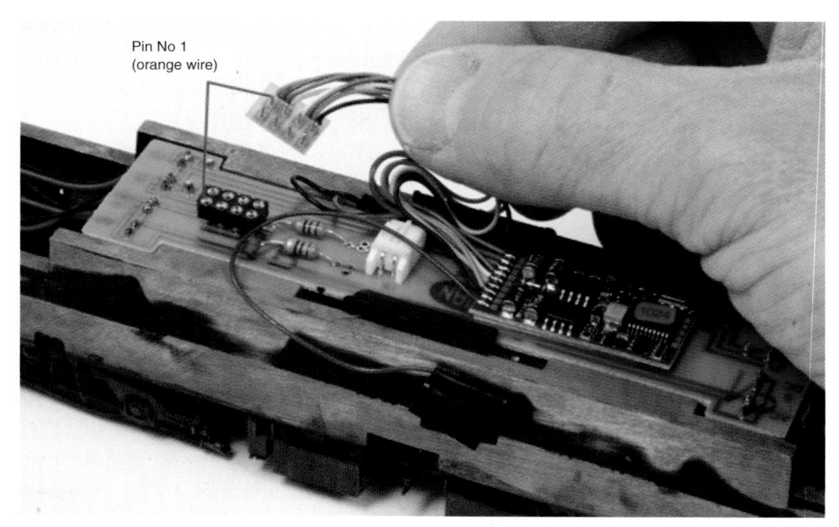


*When converting so-called DCC-ready locomotives, ensure that you buy decoders with a NEM plug factory-fitted to the wiring harness. Align the pin connected to the orange wire so it's plugged into socket No. 1, which is usually identified on the circuit board.*

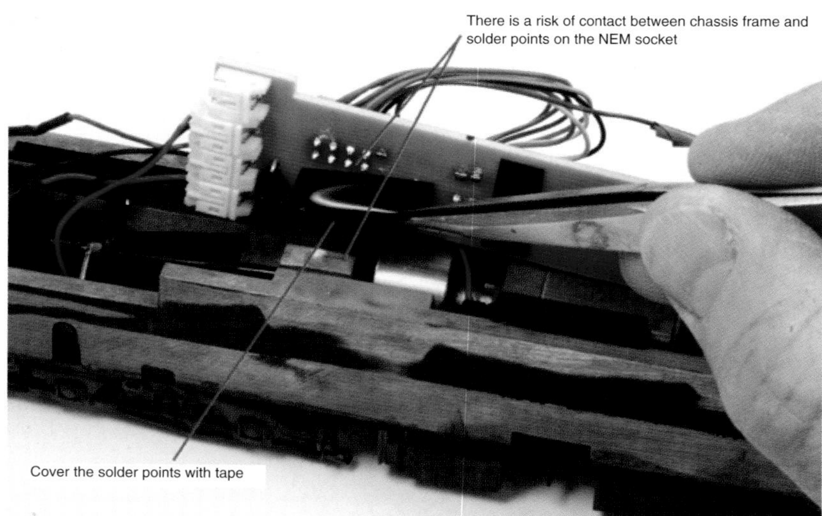


*Some soldered connections between circuit board and socket are quite large and can make contact with the metal chassis frame. The Heljan Class 47 and Class 33 models are very prone to this and insulation tape should be applied to the exposed solder points.*

shown in the photographs. It takes a second to do but will save you cooking a decoder.

Some locomotives do not have NEM/NMRA sockets. The decoder should be hardwired to the model by soldering the harness wires directly to motor terminals and power pickups. Remove all other wiring between the motor and pickups.

## ELECTRO-MAGNETIC COMPATIBILITY (EMC) DEVICES

Suppression is the process of adding components that dampen the unwanted noise signals. This is normally achieved by the judicious addition of capacitors and inductors in appropriate places. In a car, the alternator is suppressed with a large capacitor across its terminals, to stop noise being picked up by the car radio. We must do exactly the same with our motors.

One problem experienced by DCC enthusiasts is the phenomenon of runaway locomotives operating on a DCC layout. Anecdotal evidence suggests that this is caused by distortion of the digital ac signal by EMC devices, or more precisely, TV and radio suppression circuitry. Suppression is achieved by adding components that dampen unwanted noise signals and is normally achieved by the judicious addition of capacitors and inductors in appropriate places such as motor terminals in model locomotives. It is a legal obligation for the manufacturers to add them to their products, even if there is a DCC socket included on the circuit board.

While the cure would appear to be simple (remove them), a simmering debate regarding the removal of EMC devices has continued in the DCC community for years. DCC manufacturers will not suggest their removal (even though most decoders have their own suppression circuitry), not even from models fitted with DCC NEM/NMRA sockets, because they still have the potential to cause problems. I will not make a case for or against their removal – that has to be a decision for you to make for yourselves. I would suggest asking the manufacturer if its particular type of decoder will cope with EMC devices.

## CONVERTING LOCOMOTIVES WITH SPLIT FRAME CHASSIS

DCC, and simple decoder installation, was not given much thought when the split frame chassis design came about. Dismantle a split frame chassis and you soon see why they were once seen as desirable in OO gauge. Split chassis design simplifies matters for manufacturers, removing the need for complex electrical pickup devices, circuitry and motor connections. The method is extensively used in N gauge (N scale) models even today, although new models are designed with DCC in mind.

*N gauge models are particularly challenging to convert to DCC because of restricted space inside the model and the common use of split frame chassis. It's interesting to note that many N gauge models do not have EMC devices fitted across the motor terminals or the split frames.*

To explain, a split frame chassis is composed of two die-cast halves, which are a basic mirror image of each other. The halves are electrically isolated from each other with an insulating sleeve or a series of insulating bushes. Split axles complete the electrical isolation between the two sides of a split frame chassis, the only connection between the two being the circuit board and/or motor terminals. Electrical current from the track is collected via the wheels and direct contact between the axle and the bearing surface, which is usually part of the split chassis frame shell.

The motor and mechanism is accommodated between the die-cast chassis shells and a direct contact is made between the chassis shells and the motor contacts or a circuit board to complete the electrical circuit. Both sides of the split frame chassis are live: the insulating material between the halves and motor frame prevents a short circuit. On diesel-electric locomotive models, particularly in N gauge, split frame chassis design simplifies the collection of current from the bogies and supply to the motor.

N scale models produced today to 1:160 scale from manufacturers such as Kato and Atlas, and in 1:148 scale by Bachmann and Dapol, employ split frame chassis design with a circuit board clipped to the top of the chassis frame. Connections are made on both sides of the split frame insulating bush via the circuit board. Kato and Atlas models can be converted to DCC by exchanging the circuit board for a similarly shaped decoder designed specifically for that model. US-outline modellers benefit from an extensive range of bespoke decoders from Digitrax and MRC. Those older US-outline N scale models filled with die-cast metal for maximum tractive effort can be modified with milled frames from Aztec Manufacturing of Carson City, Nevada, to make room for decoder installation.

*Three different approaches to decoder installation in US-outline locomotives: (1) a Kato AC4400CW chassis fitted with a Digitrax bespoke decoder, which is designed to replace the original circuit board supplied with the model; (2) chassis from an Atlas Dash 8-40BW equipped with a factory-installed Lenz decoder, which also acts as the lighting circuit board; (3) an Atlas SD60M chassis fitted with a Lenz micro decoder of 0.5 amp rating hardwired to the factory-installed lighting circuit board.*

*Split frame chassis construction is also common in some British outline OO gauge models. This is an older Mainline (Palitoy) Class 03 shunter. It is remarkable in that the gap between the two chassis frames is wide enough to accept a single-sided, four-function decoder. It's not always that easy!*

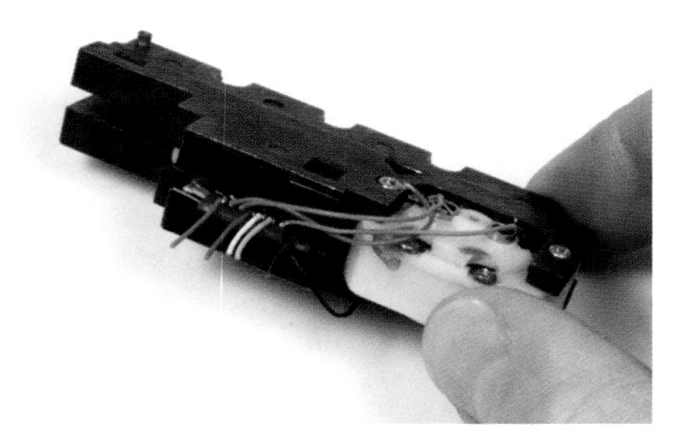

*Insulation tape was applied to the inner surfaces of the Class 03 chassis frame to ensure that the decoder was safe from accidental contact with metal. The unused function wires are cut short and can be insulated with heat-shrink sleeve or insulation tape.*

*Older steam locomotive models are usually crammed full of die-cast metal to provide as much tractive effort as possible. When they were designed, DCC either did not exist as a mainstream control system or was a fledgling technology, accommodation of which was low on the designer's priorities.*

Many older OO gauge (1:76 scale) models produced by Mainline (Palitoy) and Bachmann have split frame chassis and they are particularly challenging when it comes to decoder installation. Sometimes it is possible to file a small amount of the frame away to squeeze your favourite decoder in somewhere. The motor current draw is also sometimes within the 0.5A rating of an N gauge micro decoder and space can be found in a narrow steam locomotive boiler. Occasionally, and only very occasionally, you have to admit defeat and place the decoder in the loco tender, which is far from ideal, but at least it gets the engine running.

*There's very little room for a decoder of any type in here! That older type of motor probably draws quite a lot of current as well and so the use of a micro decoder rated at 0.5A may not be a viable option.*

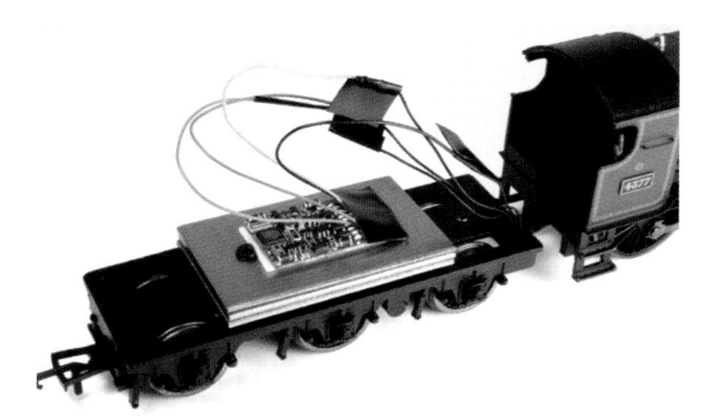

*Tender steam locomotives rarely have much to fill the tender – no coal or water required here. Tenders are a great place to locate a decoder if you are prepared to put up with the inconvenience of wires running between the tender and the locomotive. In this case, the wires that connect the tender and locomotive together have black-coloured insulation to reduce their impact.*

*The wiring connection between locomotive and tender is not obvious if wire with black insulation is used.*

## DEALING WITH MULTIPLE UNIT STOCK

Multiple units do not require more than a single decoder fitted to the driven car unless lighting is involved. Lighting can complicate matters and increase the cost. Ideally a mobile decoder should be fitted to the powered car and function-only mobile decoders fitted to unpowered trailers that will need power and control for running lights. In our example project below, the Bachmann Class 108 two-car DMU is equipped with lights. It needs only two decoders, one of which solely operates the running lights in the unpowered trailer. The function-only decoder used in trailer cars can be a cheap one as long as it has sufficient outputs for the lighting circuits. Alternatively, use an older mobile decoder that has been displaced from the front-line locomotive fleet in favour of more sophisticated decoders.

When your base model does not come with factory-installed lighting circuits, DCC offers the chance to fit running lights as an enhancement, and why not? One of the jobs you have to do is fit current collection to the bogies and factor in the purchase of metal wheel sets suitable for current collection.

When you start to look at fitting decoders to multiple units with three or more trailers, it can become longwinded. Take the Bachmann three-car Class 166 Thames Turbo, for example. Bachmann OO gauge three-car Class 158, 159, 166 and 170 units have the drive located in the centre car, as does the four-car Class 220 Voyager unit. N gauge Graham Farish three-car units suffer the same problem (although many of these models are not equipped with lights, there is room to include running lights with small SMT LEDs).

The outer cars of these OO gauge models are fitted with running lights and current collection pickups, but are unpowered. Unless the drive is relocated to one of the outer cars, three decoders are required – a mobile function-only decoder in each outer car and a mobile decoder for the driven centre car. However, there are other options depending on your available modelling funds, time and personal preferences. Here are a few ideas on dealing with multiple unit stock:

- Install a decoder in the powered vehicle to supply DCC controlled power to the motor in the same manner as you would expect for a locomotive. Additional wires could be run from the powered vehicle's decoder function wires through to the unpowered vehicles to provide juice for the running lights (clearly impractical in N gauge). If your model is to be run on a permanent 'home' layout, you could live with the practical difficulty of 'hard wiring' your multiple unit vehicles together. This would be a low-cost option, especially if there is no requirement to remove the model from the layout on a regular basis except for maintenance and wheel cleaning.
- The second option is similar to the first, except for the addition of miniature plug and socket assemblies to connect interconnecting cables between the powered and unpowered vehicles of the model. This option would enable the uncoupling of an individual vehicle from the set so that it can be removed from the layout without difficulty. The cables, together with the plug and socket assembly, could be hidden within the inter-unit gangways. The cost of this option is reasonable, although some skill with a soldering iron, together with additional time at the workbench, will be required to make the interconnecting cable, plug and socket assemblies. Furthermore, the assemblies may restrict the movement of individual vehicles relative to each other when operated on sharp curves and crossovers.
- For those modellers in need of total flexibility with their models, the only viable option is to purchase a decoder for each vehicle, eliminating the need for cable, plug and socket assemblies. This can be an expensive option, although the price of decoders continues to fall as DCC becomes more mainstream. Both unpowered outer vehicles can be equipped with decoders recovered from other models as part of a decoder upgrade programme. By 'cascading' older decoders, you can offset the cost. Failing that, purchase the cheapest decoders you can find to equip the unpowered cars and select those that have a minimum of two functions. The technical specification of the decoder drive electronics is not important as long as it has sufficient power for the running light circuits.
- When it comes to operating a DCC multiple unit, it is desirable that all the decoders installed in the train will work together. When the decoder installation is complete, all of the cars fitted with decoders should be placed on the service track at the same time so that they will take up the same address as the mobile drive decoder in one hit. This means that all of the trailer cars fitted with running lights will respond correctly to just one address and set of commands, as if the multiple unit is a single locomotive.

Some command stations will not always acknowledge the presence of a mobile function-only decoder in a trailer vehicle if it is connected to low current consumption circuits, such as those using LEDs. In that situation it is useful to place a second vehicle with a decoder connected

*One method of saving the cost of mobile function-only decoders is to hard wire multiple units together so that the running lights are supplied with power. Another option shown here is to fit miniature plug and socket assemblies. Both solutions are cumbersome and hard wiring for stock used on portable layouts is truly impractical.*

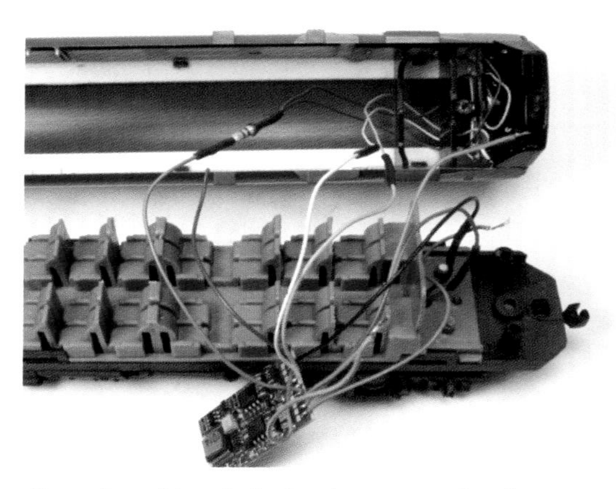

*Cascading old and obsolete locomotive decoders to function-only status can offset the cost of providing lighting controls in multiple units. This is an older Lenz LE1014 four-function decoder that has been superseded by more sophisticated versions in the locomotive fleet. Yet it still works and can control lighting circuits such as those fitted to a Bachmann Class 159.*

to a motor or similar device that has a higher load so the command station can detect that there is an installation that has been correctly made. Programme both models together. In this situation I sometimes, resort to using a completely different locomotive for joint programming with trailer vehicles. When programming is complete I reprogram the locomotive back to its original address settings.

## INSTALLATION PROJECTS

The following projects cover the majority of issues that will be encountered with off-the-shelf models.

## GRAHAM FARISH CLASS 44 (1:148 SCALE) SPLIT FRAME CHASSIS

N gauge and N scale locomotives are generally fitted with split frame chassis in which both sides of the chassis are live. Short circuits are prevented by the use of an insulating sleeve or bushes between the chassis halves. The Graham Farish 1:148 scale Class 44 locomotive is an excellent example of a model filled with weight for tractive effort but little room for a decoder, in common with so many N gauge locomotives produced for the British market. It is not equipped with a NEM 651 socket or circuit board with solder pads for a hard wire connection. A micro decoder such as a Lenz Silver Mini is a good choice for this model. It must be 'hard wired' to the motor terminals and the live chassis frame.

*The Graham Farish N gauge Class 44 'Peak' locomotive represents one type of diesel-electric locomotive introduced as part of the British Railways modernization plan 'pilot scheme'. In common with most N gauge models, it is crammed full of die-cast metal in the form of a split frame chassis.*

*There are no retaining screws that hold the chassis and body together. Simply tease the chassis away from the body as seen here. This is the most common way of dismantling N gauge and N scale models from both the UK and the USA.*

*As you might expect, there's a lot of metal in that model. The motor is captured between the frame halves. Also noteworthy is the lack of a circuit board or NEM 651 socket into which a decoder may be connected. The chosen decoder will have to be hard wired to the motor terminals, together with one track current wire to each side of the chassis.*

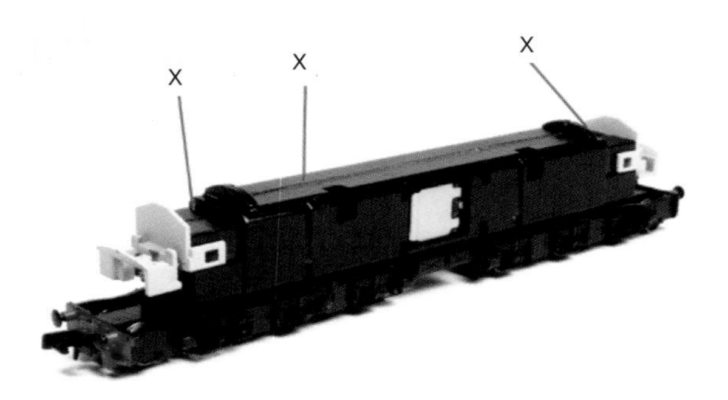


*X marks three possible locations for a micro decoder in this model. It's not looking hopeful though ...*

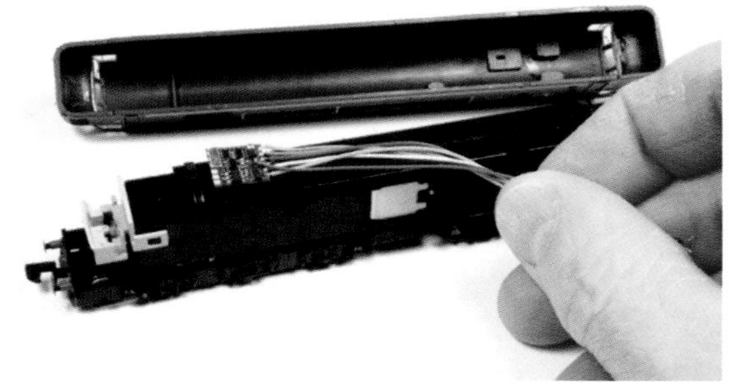

*There's one way to find out how much room we really have to play with and that is to do some test fitting with the chosen decoder.*

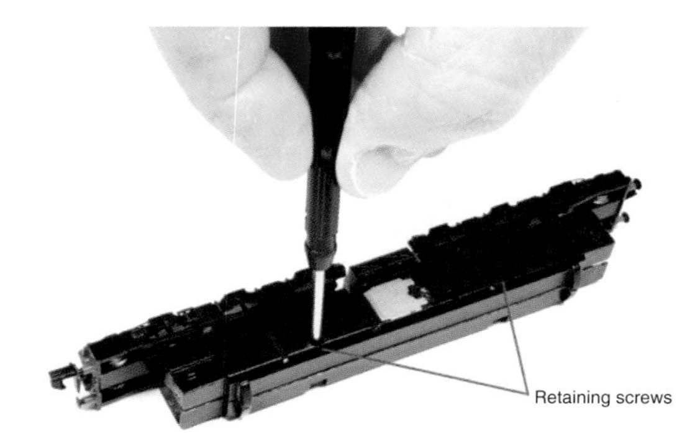


*Before taking a hacksaw and file to the chassis, it should be dismantled to remove delicate components such as gears, the motor and insulating bushes, things you don't want to contaminate with metal filings. Two screws hold the chassis frame together.*

*One useful technique to assess how much room is available in the roof of the model is to hold the body against the chassis, as here. It is becoming clear that there is no room for even the smallest micro decoder of 0.5A rating. Some cutting and filing along the top of the chassis will be required to make some room.*

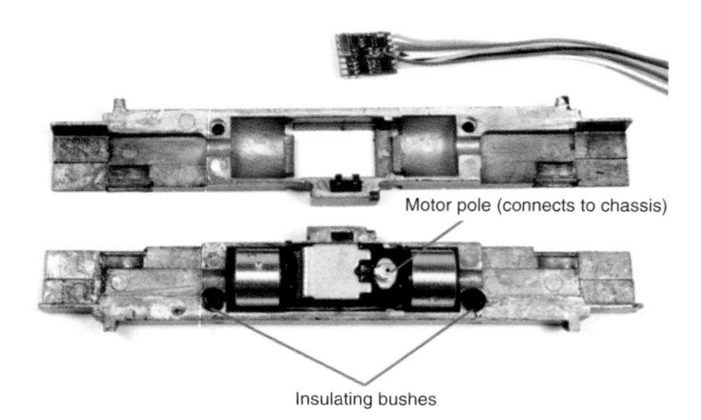


*When the frame is dismantled, you are left with two (for all practical purposes) mirror-image die-cast shells. The bogies and drive components have been removed and placed in a safe place.*

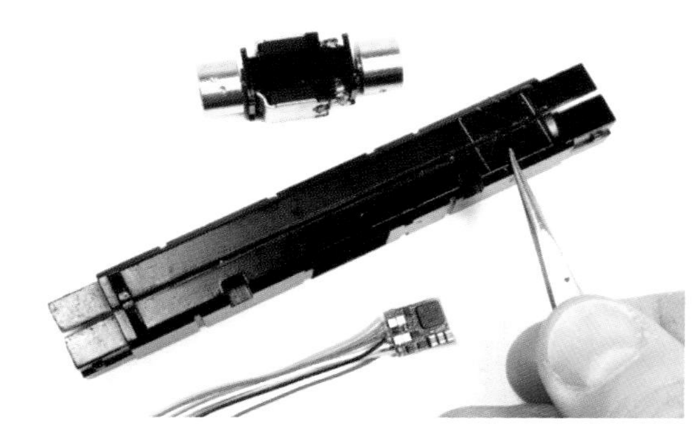

*That part of the chassis to be modified with saw and file has been identified and marked in place.*

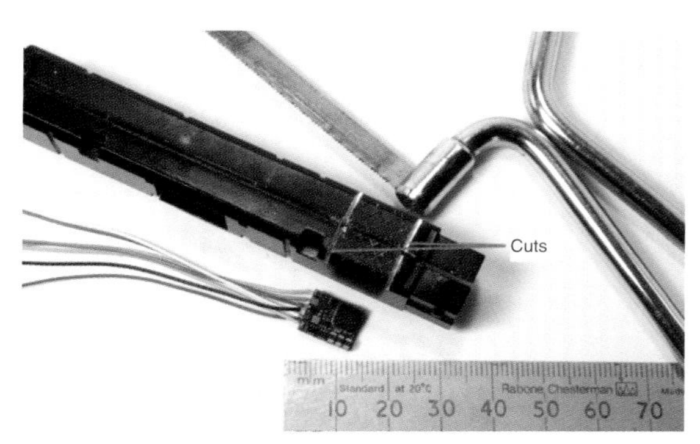


*Two cuts of 3mm depth are made with a new hacksaw blade. Both cuts are across the width of the chassis and this is done with the two chassis halves firmly clamped together.*

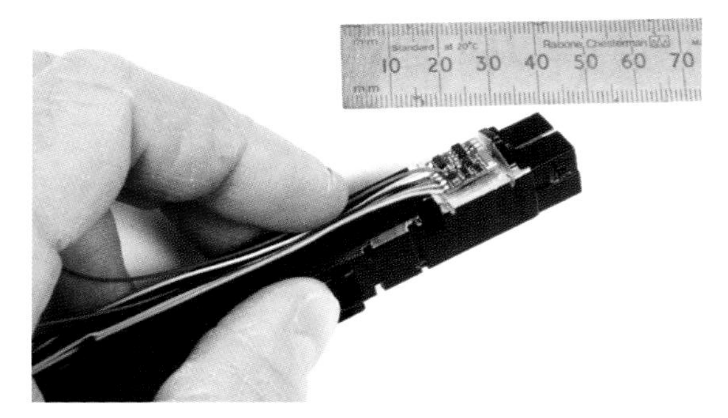


*The metal between the hacksaw blade cuts is filed away to a depth of about 3mm. The decoder is used to check the depth of the filed area. There's little point in removing more metal than is really necessary – you need every gram of weight for traction.*

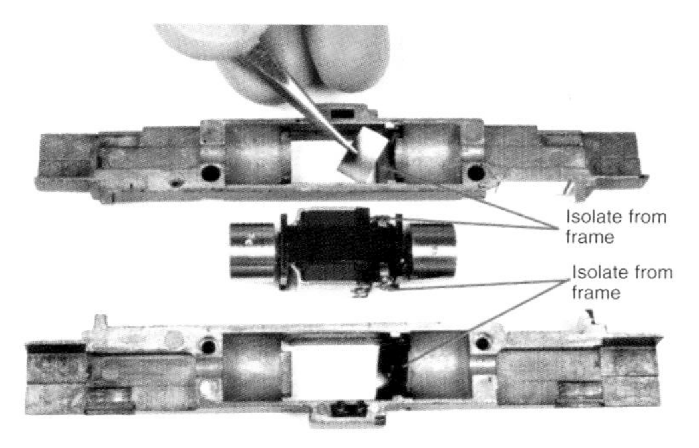


*The next task is to isolate the motor terminals from the chassis itself. This is done by applying small pieces of insulation tape to the contact point on the inside of the chassis shells. The orange and grey decoder wires will be soldered to the motor terminals, so do not apply insulation tape to them.*

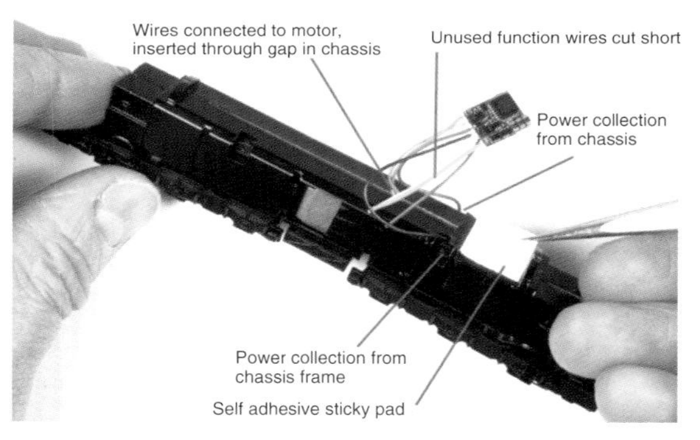


*The chassis is reassembled and the decoder fitted using double-sided adhesive tape. The black and red decoder wires are attached, one to each side of the chassis frame. The easiest way to do this is to trap them between the chassis securing screws.*

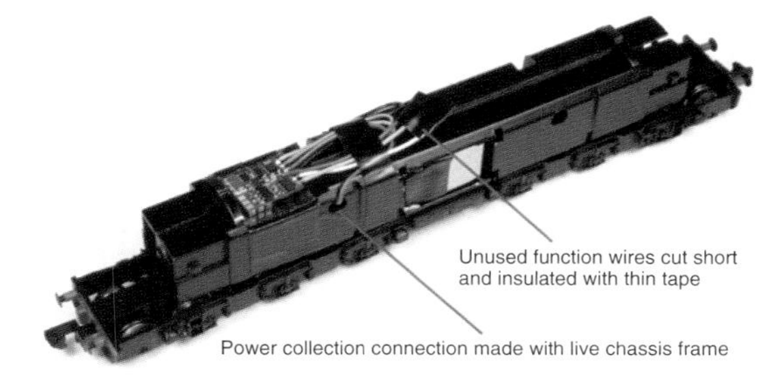


*That's the job done. Before refitting the body, place the locomotive chassis onto the service track and test the decoder installation. There may be a chance of an electrical connection between the motor terminals and the chassis frame that is bypassing the decoder. The low power in the service track is not sufficient to damage the decoder but will be able to detect a short circuit or wiring error.*

## HORNBY M7 IN 1:76 SCALE – MINIMUM SPACE 'HARD WIRE' PROJECT

When Hornby released its M7 locomotive, it was widely acclaimed for its high fidelity to the prototype together with fine detail. There was little room inside the bodyshell to accommodate the typical HO/OO 1 amp decoder, however, even though Hornby did provide a NEM 652 socket. The irony (pointed out by many modellers) is that a micro decoder would fit the model, but the typical micro decoders intended for N gauge (N scale) models rarely come equipped with NEM 652 plugs. All became clear when Hornby released its basic four-function decoder, which was small enough to fit this model, amongst others. However, that decoder does not have the functionality of more expensive and sophisticated decoders from such as Lenz, Digitrax or Zimo. The only option would have been to hard wire the larger decoders in place after removing the circuit board and socket to make more room. Even then the fit would be eye-wateringly tight.

Before attempting such a conversion using extreme measures to create room in a locomotive that appears to have insufficient space for a typical decoder, it is worth looking a little more closely because sometimes there is a simple solution that is not always immediately apparent. In the case of the M7, the answer lay in the side water tanks. Each one contains a ballast weight secured to the body with a sticky pad and a small screw. Remove one of the weights and there's room for a decoder and harness wires.

*One of the nicest UK-outline models around is the Hornby M7 tank locomotive. The level of detail is superb and it runs beautifully.*

Note that one of the chassis retaining screws is hidden in the small tank behind the front buffer beam. Hornby's instruction leaflet clearly shows how to dismantle the model and so it is worth taking a few minutes to read it.

*Oh dear, it was too good to be true! There's no room for a decent decoder in here – at least, not on first inspection.*

*What's this? A ballast weight! Can it be removed?*

*Pop one of the tank covers off the model to reveal a small retaining screw which should be released with a cross-head screw driver.*

*Tease the weight away from the sticky pad that holds it to the inside of the water tank. Suddenly there's enough room for a decent 1 amp decoder.*

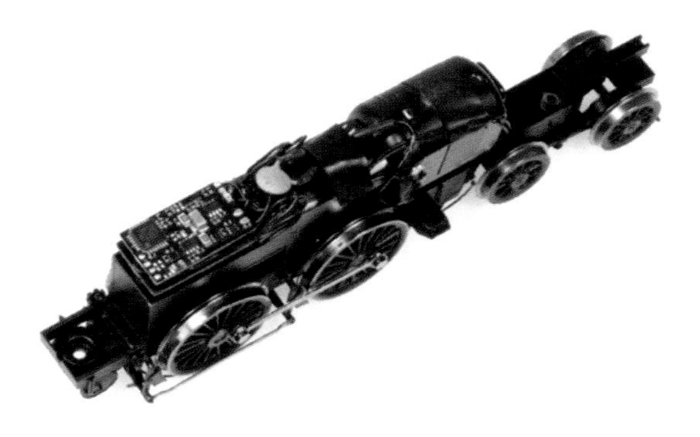

*There is the option of retaining the NEM 652 socket and plugging the decoder into that if there is sufficient room for the plug and wiring harness.*

*This shows the test-fitting of a single-sided decoder. There is insufficient room for a double-sided decoder or one with a JST socket unless it is particularly small.*

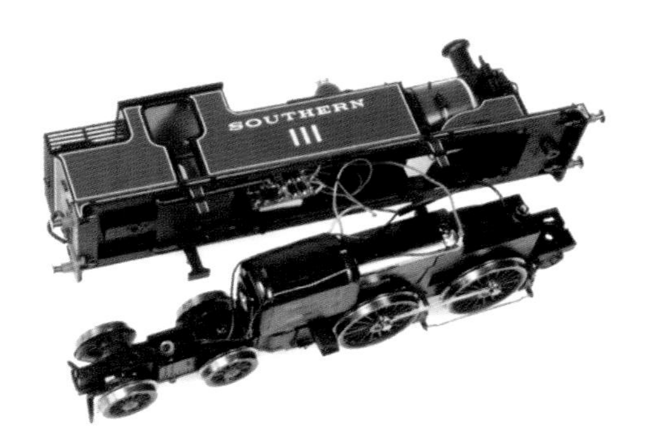


*The connections are completed by soldering the orange and grey wires to the motor terminals. If the locomotive runs 'backwards' when compared to other DCC-ready locomotives, swap the orange and grey wires around. Black and red wires are connected to the power pick up wires from the main driving wheels and bogie wheels.*

## BACHMANN CLASS 108 DMU IN 1:76 SCALE – THE CHALLENGE OF MULTIPLE UNIT STOCK

Bachmann fitted its Class 108 DMU model with NEM 652 sockets, making the conversion of this model to DCC very straightforward indeed. However, because it is a two-car multiple unit, two decoders are required. The powered car was fitted with a mobile locomotive decoder that provides power to the motor. I selected a cheap decoder for the trailer car. As long as it has sufficient function output to operate the lighting circuits, the quality of the drive output is irrelevant. The nice thing about multiple units is that there is plenty of room to accommodate the decoders. The challenge is to prevent wires from showing through the passenger compartment windows and multiple units have a number of those!

*When it comes to multiple units fitted with running lights, the rule of thumb is one decoder per car. A cheap decoder could be used in the unpowered trailer to keep costs down.*

*A foam cradle is a cheap and simple way of protecting your models during handling and decoder installation. The bogies on the Bachmann Class 108 conceal the body securing screw and they should be removed.*

*There it is! The body is secured to the chassis with plastic clips as well.*

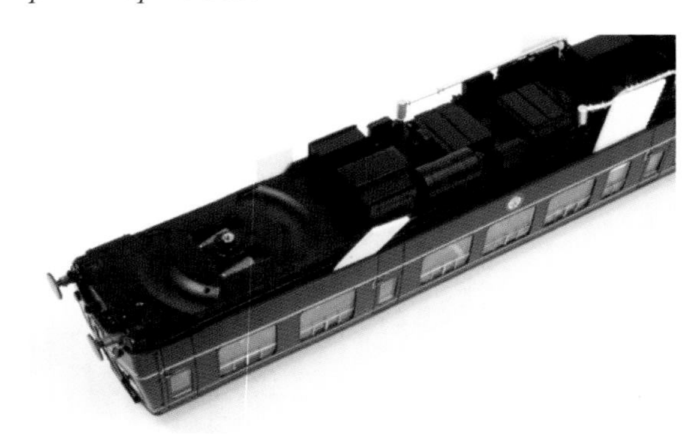

*Pieces of scrap styrene can be used to hold the body away from the chassis to ensure that all of the body-to-chassis securing clips are released. Don't force anything.*

*Once inside the bodyshell of the Bachmann Class 108 DMU, the NEM interface socket and dummy plug are easily located at the toilet compartment end of the model. This is the unpowered trailer car, which will be equipped with a low-cost decoder because all it will be required to do is control the running lights and interior lights.*

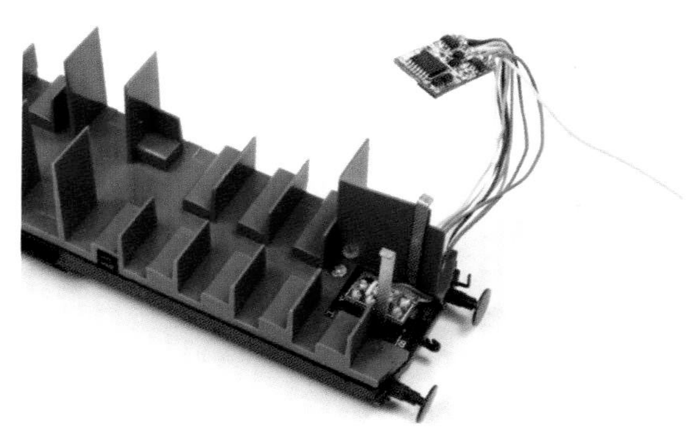

*Simply remove the dummy plug and plug in the decoder, aligning the orange wire on the plug with the No.1 pin on the socket.*

*The recess is lined with insulation tape to protect the decoder from accidental contact with metal.*

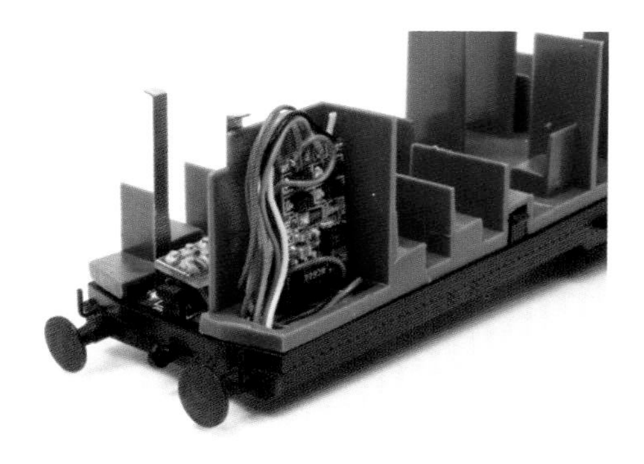

*The toilet compartment comes in very handy for concealing the decoder.*

*The decoder is plugged into the adjacent NEM DCC interface socket and firmly secured to the recess in the motor mount.*

*Turning our attention to the powered trailer, there is a recess in the top of the motor bogie mount that is a good place to accommodate a decoder. It is large enough for a typical 1 amp decoder such as the Lenz Silver.*

*The model is tested on a service track to check that the decoder installation is sound. An important point about multiple unit operation is that all cars of a multiple unit train equipped with decoders should be programmed together in one hit on the same service track so that they take up the same address and will work together as if they are a single locomotive.*

## ATHEARN 'READY TO ROLL' LOCOMOTIVES

No one can deny that the Athearn 'ready to roll' range of model locomotives represents extremely good value for money: models that could be upgraded if desired, yet have good quality mechanisms that offer very good performance. Not all of them are fitted with a DCC interface and hard wiring is the only option to install a decoder. There is a complication with this type of model that is similar to installing decoders in split frame chassis. One of the motor terminals makes a direct contact with the die-cast locomotive frame, which in turn is directly connected to the bogie (truck) current-collection devices.

One of the essentials of decoder installation is to break the direct connection between track power and motor. In this case, the motor will have to be isolated from the chassis frame and new wires installed between the bogies (trucks) and the decoder. This is how it's done.

*The basic Athearn chassis used in its 'ready to roll' range has changed very little over the years. It has been upgraded and improved to improve the quality of running demanded by today's modellers. The chassis includes incandescent light bulbs, which will need resistors to protect them from burning out when supplied with current from the decoder, and current is distributed around the chassis using crude metal clips.*

*One weakness in this chassis design is the contact between the top of the bogie and the metal clip that runs the length of the chassis. This contact, as indicated by the red arrow, can become oxidized and unreliable over time.*

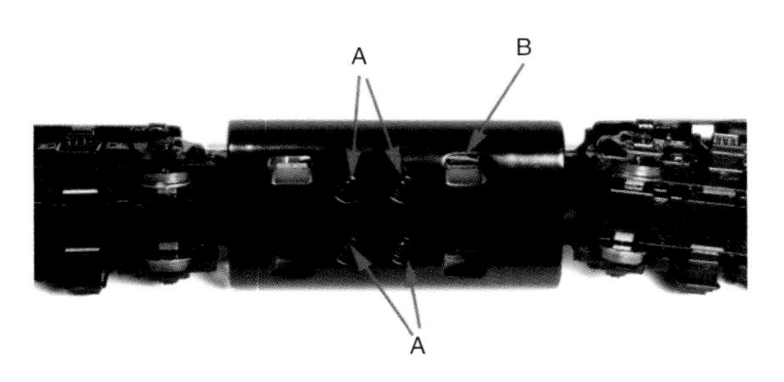


*The motor is secured to the chassis frame with four screws (A) and clips (B) located underneath the fuel tank.*

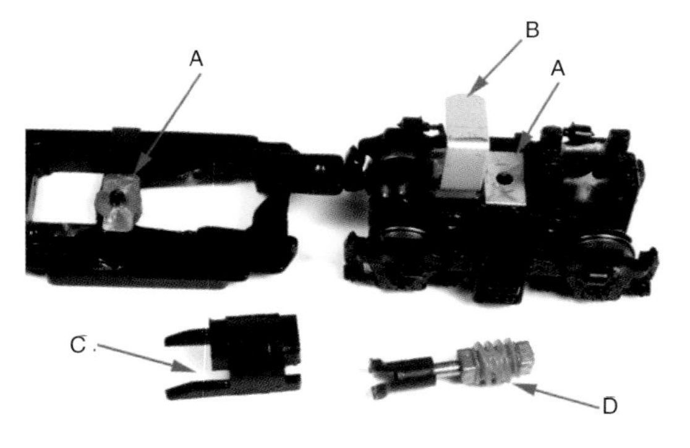


*The chassis is dismantled to gain access to certain parts of the chassis frame so that electrical contact through the frame, such as that made between the bogie pivot and frame at (A), can be isolated. The clip on top of the bogie (B) is no longer required. The bogies (trucks) are unclipped by removing the top of the gear tower (C) and gear (D).*

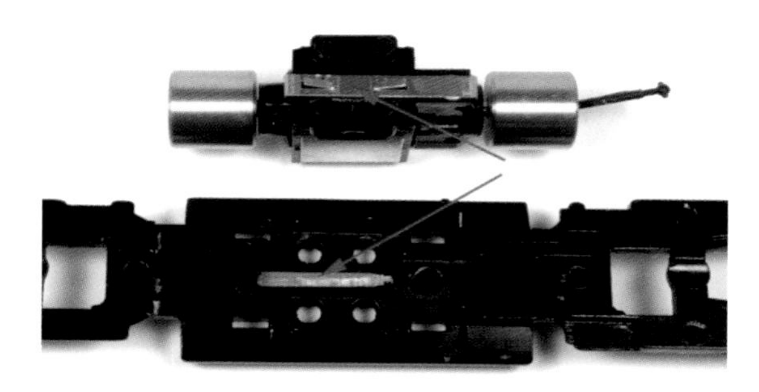

*With the chassis stripped down, it is possible to identify the contact between the motor and the chassis frame. That must be isolated with insulation tape.*

*The clip on top of the bogies is cleaned of oxidation with a file, because a new contact wire will be soldered to it.*

*This photograph shows both bogies with their new track supply wires soldered in place.*

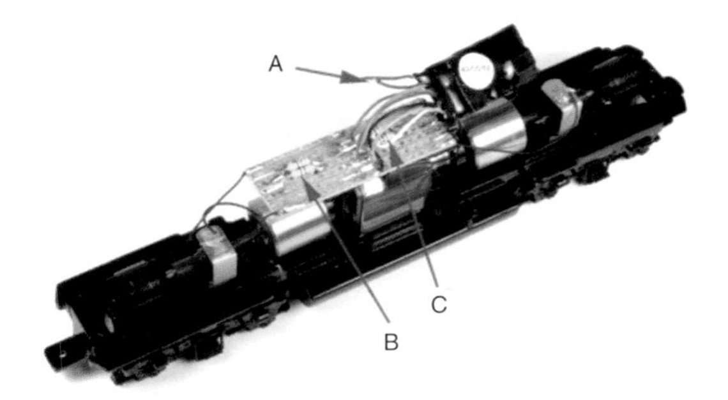


*The chassis is reassembled with insulation between the motor and the die-cast chassis frame. The track supply wires soldered to the bogies, together with leads soldered to the motor terminals, are connected to a home-made circuit board located on top of the motor. This provides all of the solder contacts for safely connecting the decoder. The model can be tested on a service track and then reassembled if all is well. A: speaker wires; B and C: lighting circuit resistors.*

## BACHMANN N2 – IN THE LOCO OR IN THE TENDER?

A very good example of a locomotive that presents particular space problems is the Bachmann N2. Will the decoder fit in the tender or the body? It all very much depends on the size of decoder you have chosen to use.

*It's a lovely looking model but there is little space in the main engine for a decoder.*

*All of the locomotive boiler and the smoke box is filled with die-cast weight, which is very securely fixed in place.*

*An alternative method would be to choose a micro decoder with sufficient continuous rating to suit the motor. Tests indicate that this particular motor drew less than 0.5 amps during normal operation. It may be worth risking a micro decoder in this model if it is restricted to light duties. A micro decoder would fit on the back of the motor frame after the circuit board has been removed.*

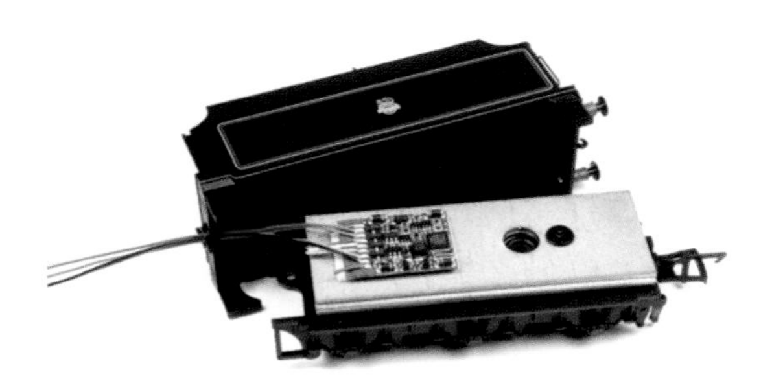

*One option is to fit the decoder in the tender, which has plenty of space in it. This involves running wires from the tender through to the locomotive, which could be a problem to some modellers.*

*The Hornby four-function decoder sits neatly on the motor frame in the space once occupied by the circuit board. There is no DCC interface socket in this model, so it would be necessary to cut the NEM plug from the decoder harness and solder the harness wires to the motor terminals and pickups instead.*

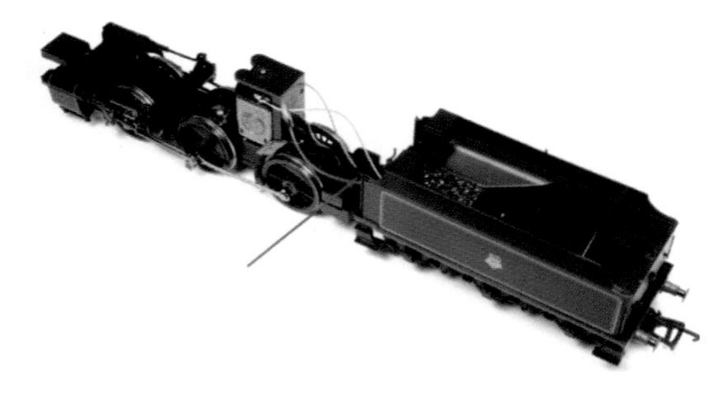

*The completed installation using the tender to accommodate the decoder.*

## US-OUTLINE N SCALE: BESPOKE DECODERS

While many older US-outline N scale models are not regarded as DCC-ready, more up-to-date models, such as the Atlas SD35, Dash8-40BW and most of the Kato range, are easily converted to DCC using bespoke decoders designed for specific models. Digitrax leads the way in N scale decoders (there are bespoke decoders available for HO scale models too), which simply replace the DC lighting circuit board by using the same chassis and motor terminals to make the appropriate connections. They are, in effect, direct plug-in and do not involve any wiring.

In common with most N scale locomotives, the chassis in Atlas and Kato N scale models are split frame, with the motor and lighting circuit board sandwiched between them. Modifying split frame chassis to make room for a more conventional decoder is beyond many modellers (sometimes that has to be done to older models and a service is offered by companies like Aztec Engineering in the USA). By replacing the dc circuit board, use is made of space otherwise not available. Note that such decoders also include replacement sunny white or yellow-glow type LEDs, which are to a better specification than the blue/white LEDs supplied with the original circuit board. Anyway, here is how such conversions are done.

*Kato N scale models employ split frame chassis for reliable operation and to squeeze all of the components into a narrow bodyshell. It has to be slackened to allow the project to be undertaken. The fuel tank unclips from the underframe and would prevent slackening of the frames unless removed.*

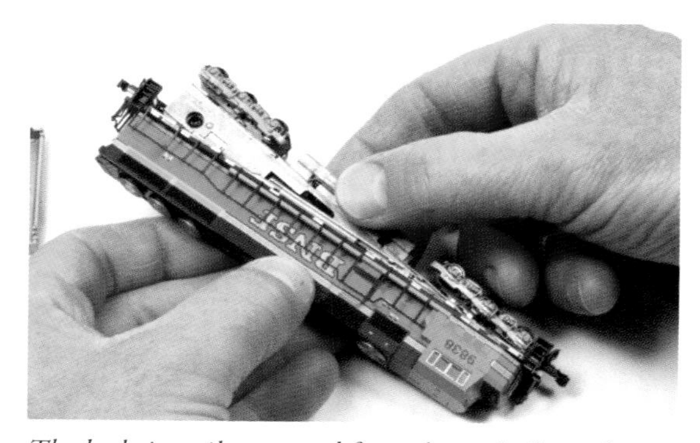

*The body is easily removed from the underframe by gently teasing it off. There are no securing screws to release.*

*Kato N scale locomotives are very highly regarded for their accuracy and quality of running. This is an SD70MAC locomotive dressed in BNSF livery.*

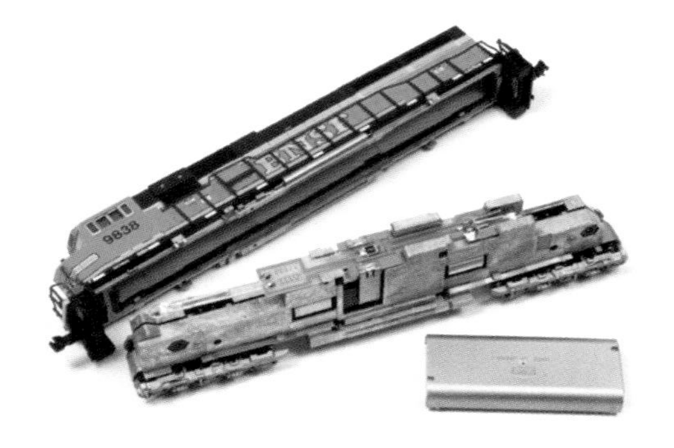

*As far as you need to go. Place the body and fuel tank mouldings to one side until later. All of the electrical components are located on the chassis.*

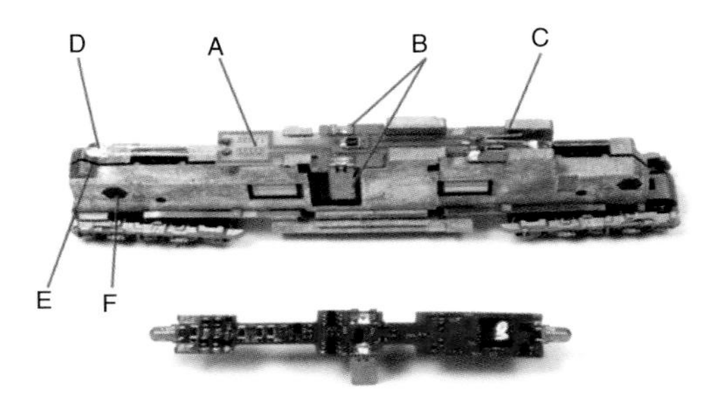


*The Kato chassis: the lighting circuit board sits on the top (A), making electrical connections with the motor at (B) and either side of the split frames at (C), white LEDs are fitted for the headlights and number boards (D); the gap between the split frames can be seen at (E) and the chassis frame securing screws are located at either end, one of which is shown at (F).*

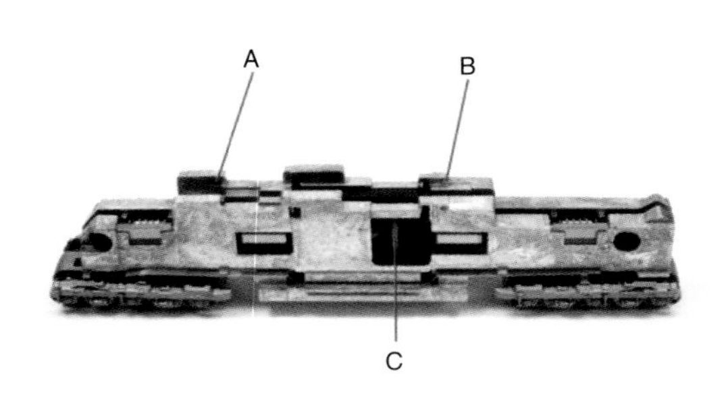


*This is what the chassis looks like after the circuit board is removed. Again, note the slot in the chassis (frame clips) for making electrical contact with the decoder at (A) and (B). The motor terminal can be seen at (C).*

*Slacken the frames by turning the screws one turn only. This just makes the removal of the lighting board a little easier.*

*Apply insulation tape to the top of the chassis adjacent to the motor mount.*

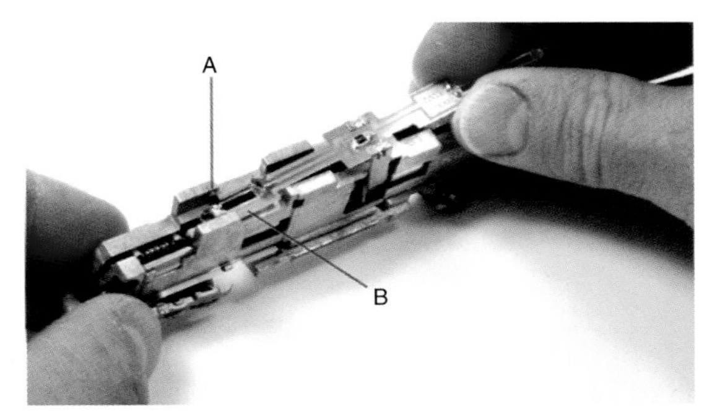


*Slide the lighting board out of the chassis, taking care not to bend the motor contacts. Note the contact slot (frame clip) in which the decoder will fit (A) and the contact surface on the lighting board (B); there will be a corresponding one on the decoder.*

*Gently slide the decoder into place, ensuring that it engages with the frame clips.*

*The completed job, with the chassis screws tightened to hold everything in place. Do not refit the body or fuel tank until the model has been checked on a service track.*

*Replace the body and fuel tank and test again before allocating a new address. That address could be the locomotive number, in this case BNSF 9838.*

*The installation is tested on a service track to ensure there are no short circuits or faults before exposing the model to full DCC track power.*

*The model may be tested on full track power and the lighting checked to see that it functions correctly. Then it is time to put it into traffic so it may earn its keep, assuming it has been correctly run-in, of course!*

*Every new decoder has a default address of 0003. A readout of the default address during testing indicates that the installation has been done correctly.*

## DCC-READY – THEY ARE NOT ALL THE SAME!

As discussed above, the term 'DCC-Ready' is sometimes misleading and no more so than when a model has no room for a decoder, or there are poor electrical connections or issues with the circuit board. The following photographs look at a selection of UK-outline OO gauge models, many of them released in the last few years. Each project, which was completed with either a 'direct' plug-in decoder or one with a wire harness, demonstrates how different they can be. All these models have in common, DCC-wise, is a NEM 652 interface socket. However, the techniques can be used to tackle any DCC-ready locomotive.

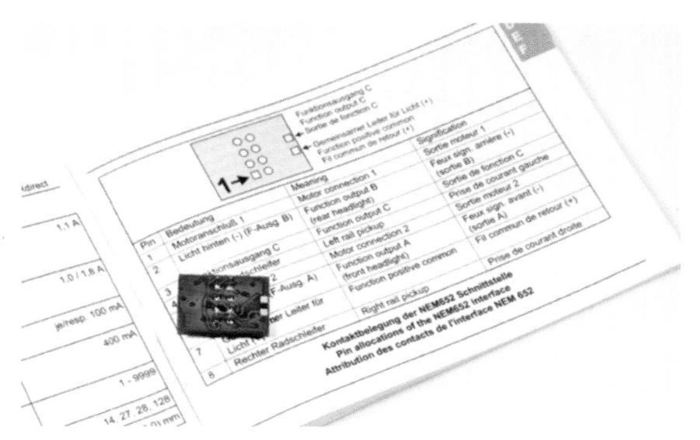

*Direct plug-in decoders, such as the Lenz Silver Direct, are growing in popularity. They simply plug into a NEM 652 8-pin DCC interface socket. Be sure to identify pin No.1 correctly!*

*A new entrant to UK-outline OO gauge models was ViTrains, which released a series of Class 37s during 2007.*

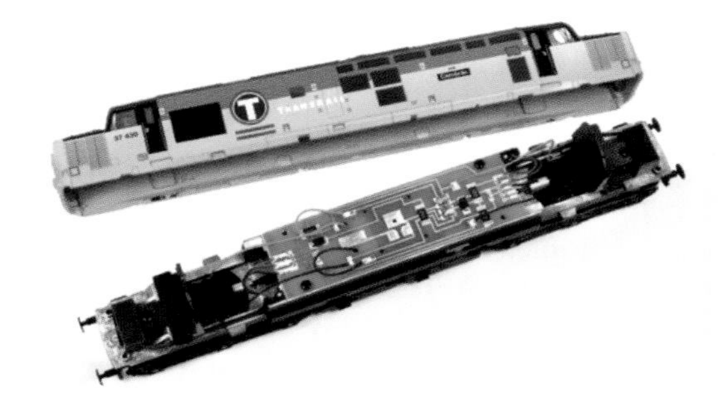

*Every manufacturer has a different take on what constitutes a 'DCC-Ready' locomotive. The powerful and smooth-running chassis supplied with the ViTrains Class 37 is shown in this picture.*

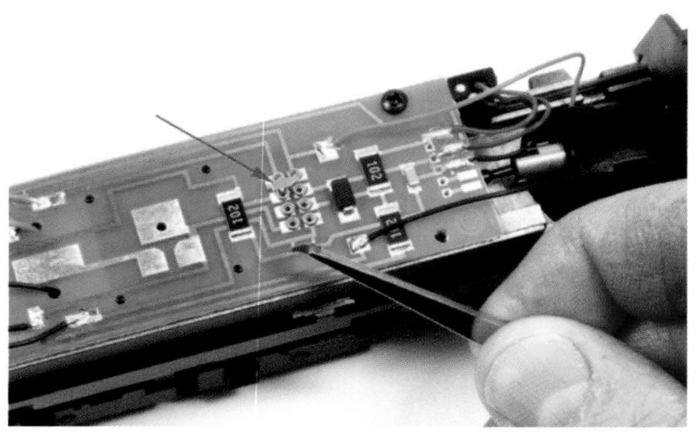

*The NEM 652 DCC interface socket is built into the circuit board. Instead of a dummy plug for analogue operation, the model is equipped with small jumper pins that have to be removed before connecting a decoder. There is no indication on the circuit board which socket accepts pin No.1 on the decoder plug, which is indicated by the arrow.*

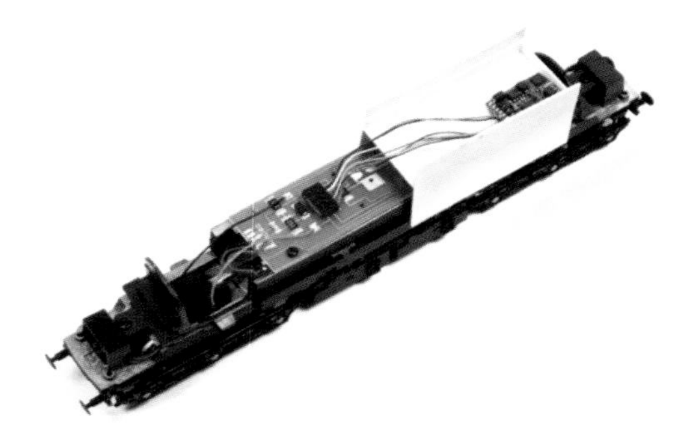

*The ViTrains Class 37 chassis with decoder fitted as indicated by the instructions. Note the use of a plastic cradle to safely hold the decoder during testing before it is secured to the model with a sticky pad.*

*Modellers were swamped with Class 37s during 2007 as Bachmann released its second version of the Class 37/4. This model was equipped with a new chassis with all-wheel drive as opposed the original versions, which were only partially driven.*

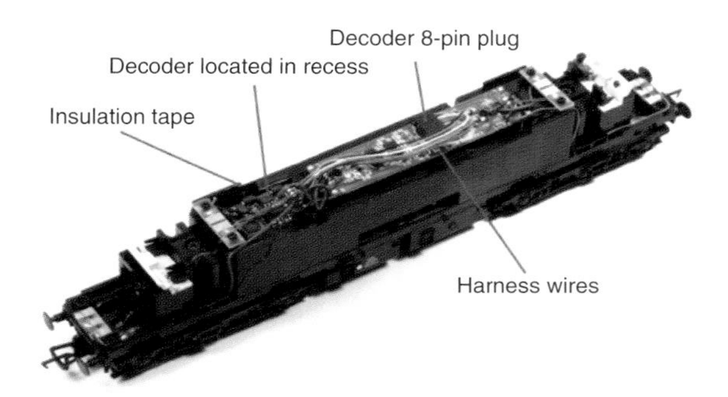


*Bachmann now provides a recess in the top of the die-cast chassis frame for accommodating a typical 1 amp decoder suitable for HO/OO gauge.*

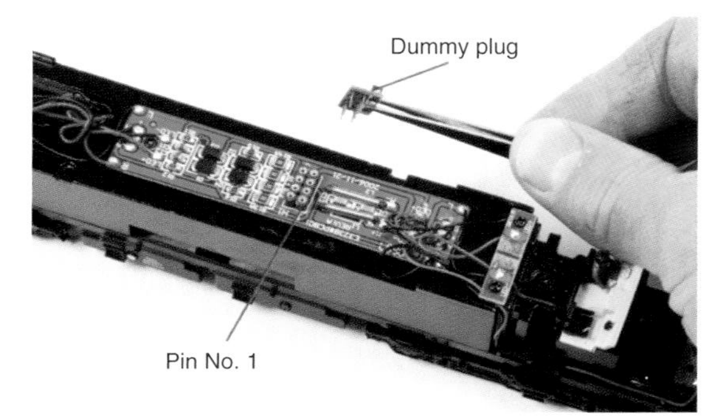


*Dummy plugs vary between manufacturers' products. Keep them safe just in case you need it to convert the model back to analogue control.*

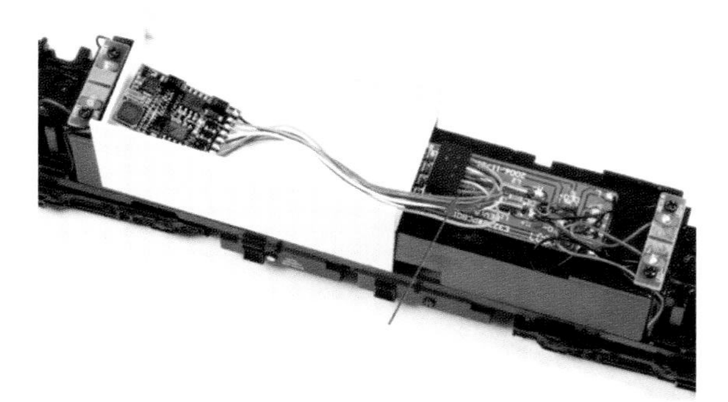

*A Lenz 'Standard' (2007) decoder was used to test installation in this model, which features cab lighting operated from a separate function button – all through the NEM interface socket. When testing a decoder loose in the model, there is always the risk of accidental contact with the metal frame, so a plastic cradle is a simple solution to preventing this from happening. Pin No.1 is indicated by the red arrow; it is also identified on the circuit board.*

*Although basic when compared to recently tooled high-tech models, the old Lima products now available under the Hornby label, such as the Class 66, are very popular. Hornby has upgraded the electronics to include a DCC interface socket.*

*There is a great deal of room for a decoder in this model. The No.1 socket position is clearly marked.*

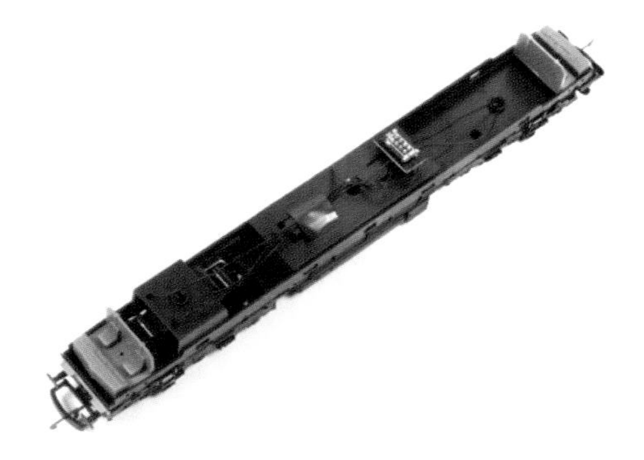

*A large decoder could be safely placed in the recess over the trailing bogie without fear of touching any metal components.*

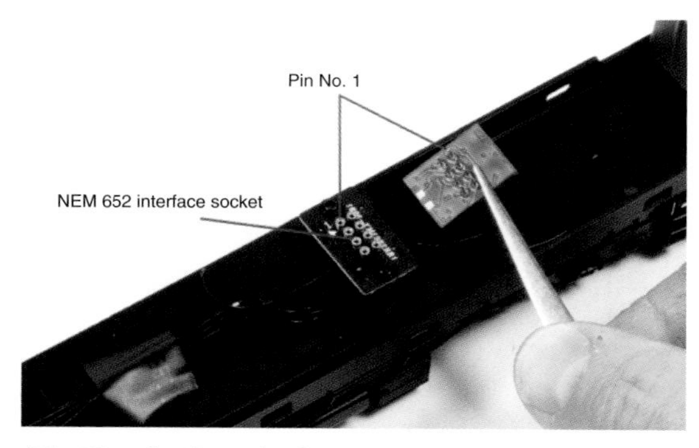


*The Hornby (Lima) Class 66 is not equipped with any lighting circuits. All the decoder will have to do is provide controlled current to the motor. A direct plug-in type is an ideal choice for this model.*

*When first released by Lima the Class 73 model was highly regarded. It is now available under the Hornby label, equipped with a new drive bogie and revised circuitry.*

*A two-second job with no need to worry about insulation tape or anything like that. Just ensure you pop pin No.1 in the correct socket or the model will run backwards relative to other similarly equipped locomotives.*

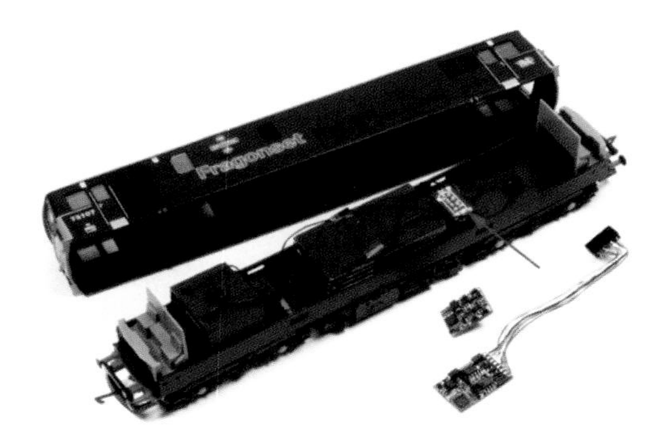

*Hornby has provided a DCC interface socket in this model, which is ideally placed to accept a direct plug-in decoder or one with a wire harness.*

*The decoder sits neatly over the DCC interface socket and is ready for testing and programming with a unique address.*

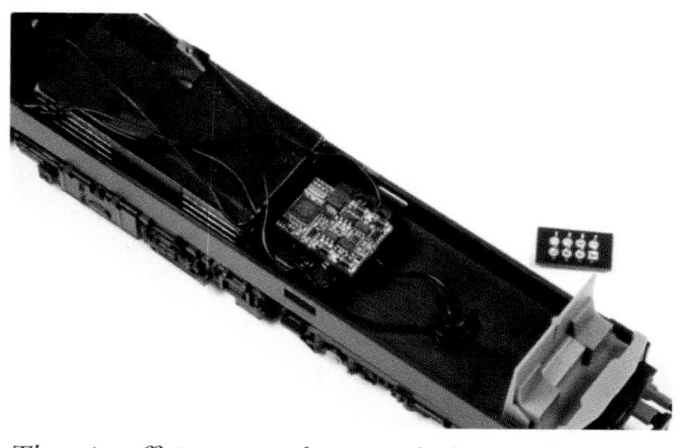

*There is sufficient room between the large ballast weight and the decoder socket to accommodate a Lenz Silver 'Direct' decoder without making unwanted contact with metal.*

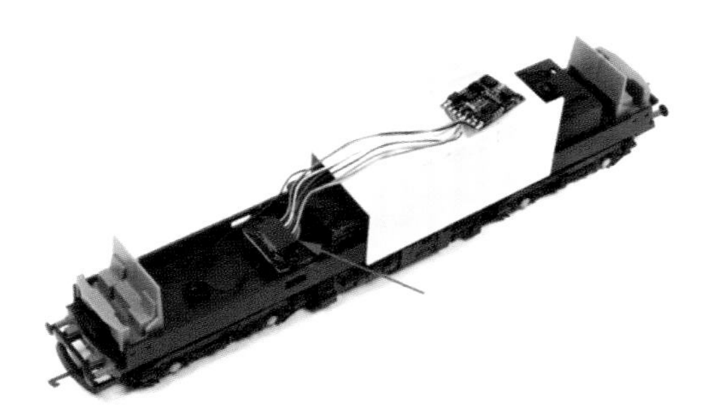

*A conventional decoder and wire harness also fits the socket. The Hornby (Lima) Class 73 has plenty of room for locating a decoder on the chassis frame. The arrow shows the position of pin No.1.*

*Highly acclaimed and DCC-Ready – the Hornby Class 50 is a new tooling of a popular class of locomotive.*

*'New-generation' Hornby models are considerably more complex in design, fitted with lighting circuits and a mechanism designed to drive a representation of a cooling fan. A DCC interface socket is provided on the circuit board.*

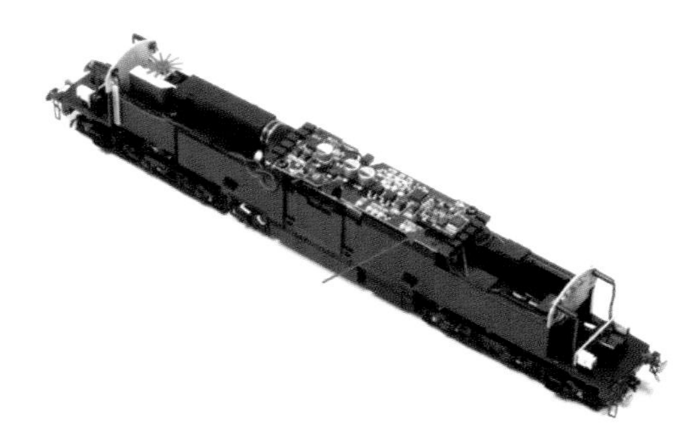

*The same Class 50 chassis fitted with a Lenz Silver 'Direct' direct plug-in decoder: a five-second job because the bodyshell of this model is simply clipped onto the chassis.*

*Another of the new-generation toolings from Hornby is the Class 60, a heavy freight engine in service from the 1990s to the present day. The Hornby model is equipped with all-wheel drive, working lights and a DCC interface socket.*

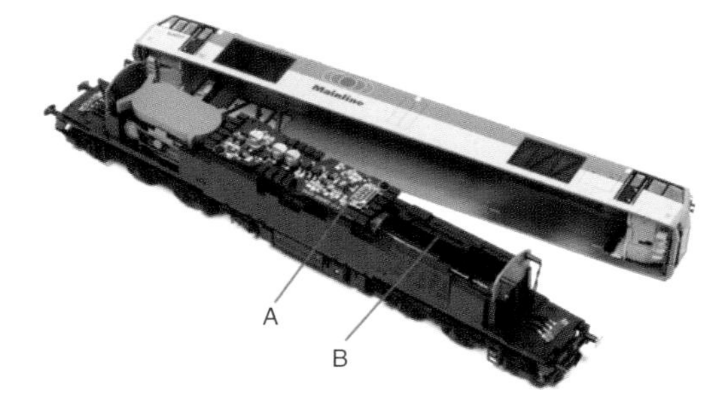


*The Class 60 shares the same advanced chassis design as the Class 50. The DCC interface socket and dummy plug are located at A. Hornby has provided a recess in the metal chassis frame that is large enough to accept a single-sided 1 amp decoder at B. The metal would have to be covered with insulation tape before a decoder is fitted. Contact between metal and the decoder must be avoided.*

*One method of providing a safe place for a decoder is to fit a simple deck of 20thou styrene card.*

*The decoder is placed in the centre of the styrene card and secured with a sticky pad. No contact is possible with the metal chassis frame.*

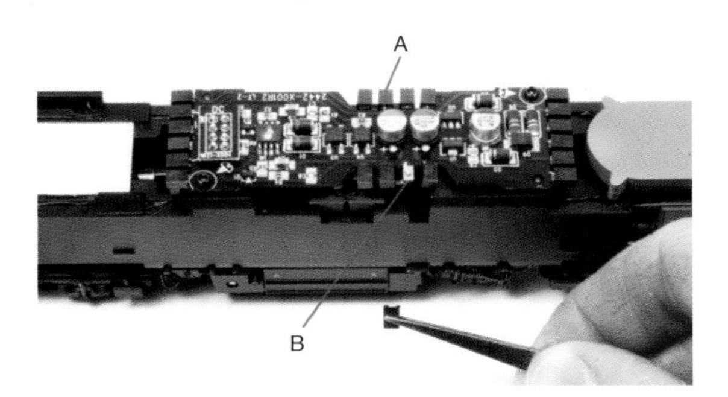


*One feature of new-generation Hornby models is the use of plastic clips (A) to secure wires to the circuit board (B). This method is not as secure as using solder and many electrical faults have been traced to poor contacts on the circuit board. It is worthwhile taking the time to solder all of the wires directly to the circuit board, retaining the plastic clips as insulation.*

*The Hornby Class 31 is another new model with many refinements in the chassis and drive department, including all-wheel drive, lighting circuits and a great deal of useful ballast weight.*

*Unfortunately, although equipped with a DCC interface socket, the Class 31 has no safe location to locate a conventional decoder and the available space between the roof of the model and circuit board is slight: a tight fit for a direct plug-in decoder.*

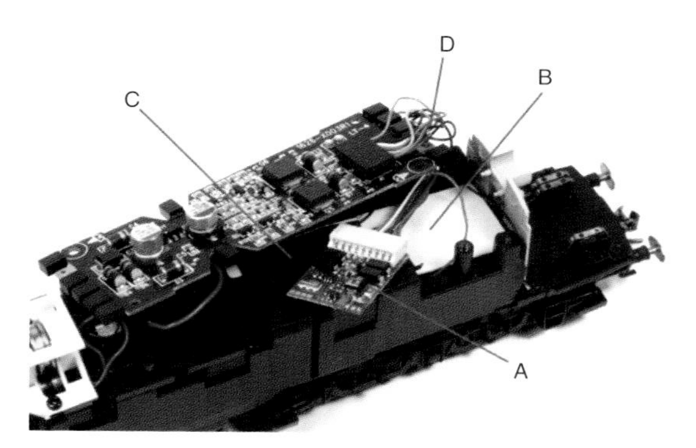


*The model will host a typical HO/OO decoder, such as this Lenz Gold (A) fitted with a JST socket. The best solution is to remove the circuit board to reveal the mechanism on one end of the chassis. That space is covered over with a piece of styrene card (B), secured with superglue. The exposed metal of the chassis frame is covered with insulation tape (C) to create a safe chamber for the decoder. The circuit board is then replaced and the wires routed around the end of the circuit board (D) to avoid snagging with the bodyshell.*

*Class 20s make great subjects for DCC layouts: they operated in pairs on various types of freight train during their heyday and are fascinating locomotives to equip with working marker lights. The class celebrated its fiftieth anniversary in 2007. The Bachmann model has seen much refinement since its introduction in 2004, including onboard sound decoders and speakers. This model is fitted with Bachmann's 21-pin decoder DCC interface plug and socket arrangement. The original releases were equipped with 8-pin NEM 652 interface sockets.*

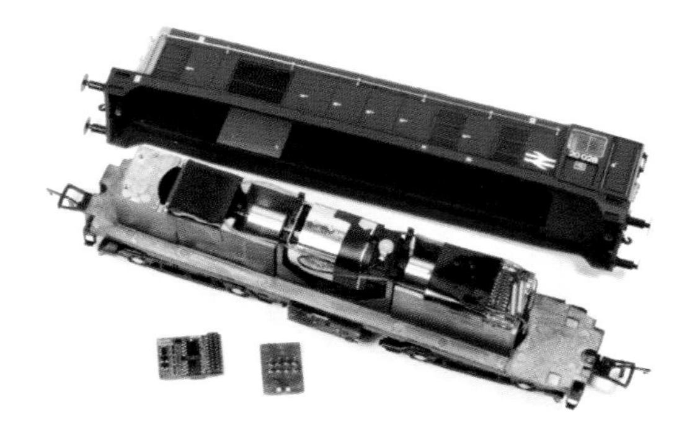

*The model has a narrow body, most of which is filled with die-cast metal for traction. The decoder socket is located at the front of the long bonnet.*

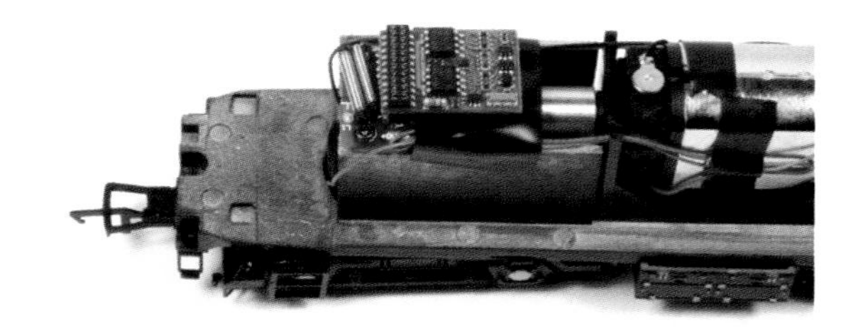

*Sufficient room exists for a 21-pin interface decoder, making the use of an adapter socket unnecessary. A piece of insulation tape has been applied to some exposed metal as insurance against contact with the decoder.*

CHAPTER 5

# Lighting for Locomotives, Rolling Stock and Multiple Units

*Inverness, 23 March 2007: darkness falls and departure of the 20.40 Inverness–Euston from the Highland capital is not far away. EWS Class 67, 67 012, was in charge and displaying night-time running lights when it was photographed. To model this would require a total of ten functions: two for tail-lights, two for the side marker lights, two for the top marker lights and four for independent control of the four headlights. It can be done with two decoders, function mapping and an advanced DCC system capable of supporting more than ten functions on the throttle.*

## INTRODUCTION

One of the benefits of DCC is constant and independently controlled lights and as part of our journey into Digital Command Control we take a look at how to fit and operate running lights and the associated lighting effects in our models. There's a great deal on light emitting diodes (LEDs) in this chapter because they are the most commonly used light sources in this type of modelling. Many modellers prefer to replace factory-installed incandescent lamps with LEDs, too. Read on to find out more about lighting techniques and prototype information:

- Running lights in full-size locomotives.
- Associated operating lights and interior lighting.
- The different types of LEDs and their advantages over incandescent lamps.
- Mobile function-only decoders and how to reuse old obsolete decoders.
- Decoder output functions.
- How to reduce decoder output voltages to protect decoders and incandescent light bulbs.
- How to wire up lighting circuits to decoders.
- Function-mapping.

At the end of the chapter there are sample projects covering most modelling situations that you can follow and adapt to your particular needs:

- Running lights in GP38-2 locomotives – ditch lights and head lights.
- Oil lamps on steam locomotives.
- Marker lights on older British diesels.
- Using lighting kits in diesel locomotives with minimal space.
- An interior lighting project on a heritage diesel multiple unit.
- Modern multiple units.
- Rolling stock lighting projects.

Lighting is not restricted to locomotive head- and tail-lights. It can be incorporated into multiple units to operate door warning lights and interior lighting, and used in rolling stock to illuminate the interior and tail-lights as well. Once you begin to think about the possibilities, you soon come to realize that the number of functions provided on the typical HO/OO decoder may be insufficient to meet your ambitions for a given model and a secondary decoder may be required, one with sufficient functions to supplement the drive decoder. Such decoders are called function-only mobile decoders and can provide as many as six additional outputs. They do not have a primary output to control a motor.

O gauge modellers can buy large decoders with more than the normal four functions, while some decoders intended for HO/OO gauge may have only three or four functions. Some, such as the Digitrax DH163, have six functions. You will need to think about the requirements of your project before deciding on a decoder or combination of decoders. Some projects using more than four outputs will require a process called 'function mapping' of additional outputs on a function-only decoder. Function mapping is simply the allocation of function control to specific buttons on the throttle. (For more detail on function mapping, *see* below.)

N gauge modellers are faced with an understandable space restriction. Installing directionally controlled running lights may be challenging and going for more than that is regarded as impractical. None the less, new-generation N gauge models are being equipped with lights and NEM 651 DCC interface connections, bringing working head- and tail-lights within reach. Typical micro decoders for N gauge are usually capable of providing power for two separate lighting circuits, which is sufficient for directionally controlled head- and tail-lights.

US-outline modellers currently enjoy a growing range of bespoke decoders, which are styled to fit individual N scale models produced by Kato, Atlas and others, and are the same size and shape as the DC circuit board. It's a straight swap – including the LEDs. Some special lighting effects, however, such as flashing ditch lights, might not be that easy to achieve without further decoder functions. Space is at a premium for decoders, yet alone more wires, but the effort

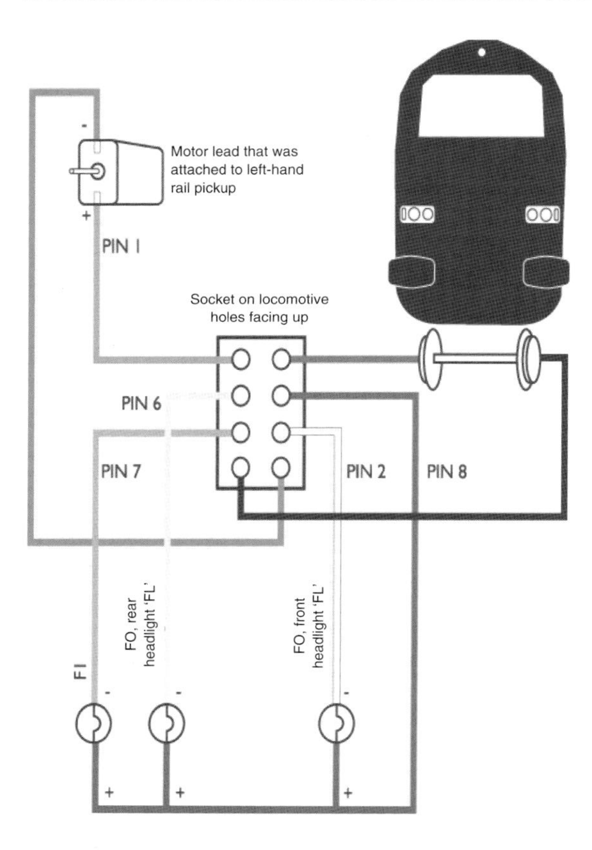


*The relationship between decoder output harness wires, the NEM 652 DCC interface plug and lighting circuits.*

is always worthwhile since working lights bring models to life, whether it be cab lights and strobes on large scale models or simple directional lights in N gauge. DCC makes constant and controllable lighting possible. This can be achieved either with individual LEDs and resistors to create a bespoke project or by relying on lighting kits.

Introducing working lights to your models also brings, for the first time, some practical modelling techniques to our DCC story. Up to now we have mostly dealt with the practical issues surrounding decoders, wiring, digital devices and busses of various kinds. When it comes to installing lighting, either for internal or for running lights, or modifying lighting circuits already provided by the model manufacturer, it is likely that some modification will have to be made to the project locomotive or rolling stock. For this you will need to have some skill with twist drills, pin vices, adhesives, fillers and making small, simple changes to the external appearance of a model, such as touch-up painting.

## LIGHTING ON FULL-SIZE TRAINS

It helps to have an idea of how the running lights on full-size trains actually work. Before embarking on your project, research your chosen subject to see how the lighting functions are used in everyday operation. The reason for this is simple: mainstream manufacturers, particularly those for the UK-outline market, install only basic lighting in their models. They may install a DCC interface socket but wire up the head- and tail-lights to two decoder outputs only (A and B), so they change when the direction switch is changed to reverse the running direction. The tail-lights remain illuminated on locomotives when hauling a train. If this were to happen on full-size railways in the UK, signalmen and other operating staff would bring the train to a stand for inspection, because that is an incorrect operating procedure. They should be turned off so that the tail-light on the last freight car or wagon correctly identifies the rear of the train. In the USA, the same role is played by the 'end of train device', although rear red tail-lights are not as common on locomotives.

This is not so critical when modelling multiple unit stock because, if more than one unit is run as one train, they can be 'consisted' together, using the consisting feature that most DCC systems offer. This also brings in different control of head- and tail-lights at the ends where the units are coupled together.

Modellers fitting lighting to modern British diesel and electric locomotives and multiple units need to be aware of the night-time lighting rule by which, in order to avoid dazzling drivers of oncoming trains, the left-hand headlight is switched off and the off-side headlight on the right is used instead. In the USA, what modellers refer to as Rule 17 lighting (you will see reference to this in decoder manuals) covers the same ground: to avoid blinding engineers of oncoming trains, Rule 17 requires that engineers must 'dim' the headlights as they approach each other as well as during certain other operating situations, such as yard work and shunting. On the full-size railroads, Rule 17 has been replaced by more comprehensive lighting regulations (Rule 5.9.1 etc.), which would be fun to create in model form with DCC.

Oil lamps on British steam locomotives offer a different set of modelling challenges. Steam locomotive lighting is (mainly) restricted to lamps on the front locomotive in positions that represent the type of train that the locomotive is hauling at the time (some steam locomotives were equipped with electric lights, which were not always used). There are none of the special effects, such as day and night lighting or independently controlled tail-lights, that are required with diesel locomotives. Unlike electric lamps, oil lamps are not simply turned on and off at the throw of a switch. Using the decoder's built-in switching feature, which automatically changes the direction of head- and tail-lights when the direction of travel is changed, is not appropriate! When a steam locomotive changes direction, the oil lamps will remain illuminated, no matter what happens, until they run out of oil. One of the projects in this chapter looks at oil lamps on stream locomotives.

The accompanying illustrations offer an idea of some of the lighting challenges the modeller faces if more authentic operational detailing is desired.

*Another Class 153 unit shows off its forward running lights, including the illuminated headlight on the left. If operating at night, the offside or right-hand headlight will be used. This photograph of No. 153333 was taken at Walsall in March 2003: (A) forward marker lights; (B) daytime headlight; (C) night-time headlight; (D) tail-lights (both are illuminated when used).*

*Multiple unit stock is easily fitted with head- and tail-lights and looks very effective. The options include tail-lights and daytime headlights linked to two functions on the decoder. They change when the reversing switch is changed. This leaves spare outputs on a four-function decoder for orange 'door lock' lights. If night-time headlights are required, the four main headlights would have to be separately controlled and the normal four-function decoder will not have enough outputs to achieve this, so a function-only decoder would be required to provide additional controlled outputs. This photograph shows a Class 153, No. 153351, stabled at Carlisle.*

*Amtrak P42 No. 89, hauling the 'GrandLuxe Express', formerly called the 'American Orient Express', leads two sister locomotives up the long climb over Marias Pass towards East Glacier on the BNSF Hi-Line in July 2007. Passenger trains in the USA are equipped with headlights, ditch lights and tail-lights. Ditch lights may be set for alternate flashing, which is set up in some types of decoder as an alternating flash between two function outputs. A model of this locomotive will probably need six independent functions, together with special lighting effects. A Digitrax DH163D, with its six functions and special function effects that can be switched on by adjusting values in the decoder's CVs, would probably be an ideal choice.*

*A modern variation on steam locomotive marker lights is seen on B1 No. 61264. A portable electric headlight, evident between the oil lamps, is painted black to minimize its impact on the cosmetic appearance of the locomotive. No. 61264 was photographed at Fort William in June 2006 on the hugely successful 'Jacobite' service that runs daily between Fort William and Mallaig as a seasonal tourist train throughout the summer months.*

*BELOW: Some types of rolling stock are equipped with running lights to save the use of portable tail-lights. Propelling Control Vehicles, once used on the rear of mail stowage trains, were equipped with lighting that was put to good use, as seen on No. 94323, which was displaying its tail-lights at Reading in November 2002.*

*Amtrak Superliner coaches are equipped with built-in tail-lights, which are displayed at the rear of the train instead of the British practice of using a portable flashing tail-light. Superliner passenger car No. 32060 was photographed at the rear of the eastbound 'Empire Builder' when observed at Essex, Montana, in July 2007. This effect would be very easy to achieve with a cheap single-function decoder and a couple of resistors and red LEDs. A model of this passenger car must be capable of collecting current from the track through its wheels to supply power and digital data packets to the decoder.*

*Modern US-outline practice includes two high-powered headlights designed to illuminate an object more than 300 yards away, together with two 'ditch lights', which are fitted either below or on top of the front pilot of the locomotive. Some locomotives, usually switchers, have ditch lights at either end and some have them only at the front (remember the letter F identifies the front of the locomotive). Furthermore, some research should be undertaken to find out which railroads use alternately flashing ditch lights, which are activated when the horn and bell are sounded, and which do not. Ditch lights were introduced to improve visibility of approaching trains to reduce accidents at grade crossings. A coal drag led by BNSF 8870, an SD70MAC, displays both its ditch- and headlights when cresting the summit at Mullan Pass in Montana.*

*More possible lighting effects are demonstrated by this Class 90 and the first coach of the Caledonian sleeper to Aberdeen, Fort William and Inverness, which was photographed at London Euston station in March 2007. Opportunities to use decoders for additional lighting effects in Mk.3 sleeper stock include door lock lights (A), interior lighting (B) and built-in tail-lights (C). It is worth noting that the latter are not always used on the Caledonian sleepers and portable flashing tail-lights may be substituted instead.*

*Portable tail-lights can be represented by one of the cosmetic details available from Springside Models, a piece of light guide and a red LED. The decoder must be capable of being programmed with a strobe flash for this type of portable flashing tail-light.*

## CASCADING OLDER DECODERS TO LIGHTING PROJECTS

Technology advances at an alarming rate and sometimes it seems that yesterday's advanced decoders are today's obsolete technology. The decoders themselves, however, will still function as they always have, because no one will have told them they are obsolete. Even if your decoder is elderly, it will still have a useful role to play on your layout. One trick that many modellers adopt is to cascade older obsolete decoders to function-only use as they upgrade their locomotive fleets with the latest versions. The motor drive function on a cascaded decoder is no longer used; the objective is to make further use of the output functions to control additional lighting effects in the locomotive. The only disadvantage of this is when the older decoder does not have programmable lighting effects that may be designed for a specific project such as Rule 17 lighting, strobe or flashing effects. This can also be complicated if the decoder is sufficiently obsolete not to support the latest function mapping, which could be essential for making a lighting project work in a specific locomotive.

When an old decoder will not support more modern practices, I usually cascade it down the fleet to power lighting systems in rolling stock where the decoder will not be required to work with any other, thereby avoiding any potential conflict with function mapping or special lighting effects. Whatever you choose to do, do not disregard and dispose of obsolete decoders because they do have a use, perhaps for introducing interesting lighting effects in rolling stock that otherwise would not be economical if a specific function-only decoder had to be purchased.

## FUNCTION-ONLY DECODERS FOR LIGHTING FUNCTIONS

Most DCC suppliers offer decoders that are specially designed to provide function outputs for onboard systems and nothing else. These decoders are meant to supplement the function outputs supplied on the primary drive decoder. For the sake of identification, function-only decoders are described as secondary decoders in this book. They are usually economically priced in terms of providing output functions, but they can make the cost of a rolling stock lighting project quite expensive if only one of the four or six provided functions is likely to be used. Under those circumstances, the modeller may wish to use a bridge rectifier to supply current to interior coach lighting, if there is no need to control it from hand throttle, or use an old obsolete locomotive decoder cascaded by the introduction of better ones.

## DECODER FUNCTIONS AND POWER CONSUMPTION

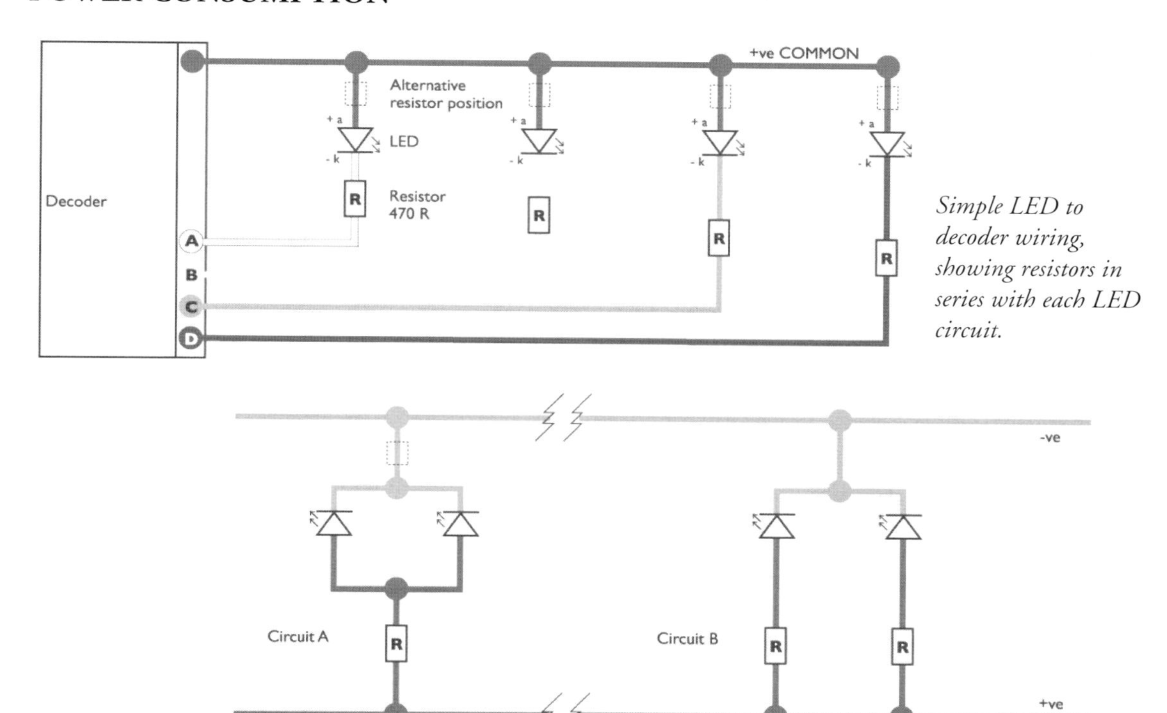


*Simple LED to decoder wiring, showing resistors in series with each LED circuit.*

ABOVE: *The best approach to wiring multiple LEDs (or bulbs) to one function output is to provide each with a separate resistor, as shown in circuit B, as opposed to one resistor for two LEDs (or bulbs), shown in circuit A – that's if space permits, of course!*

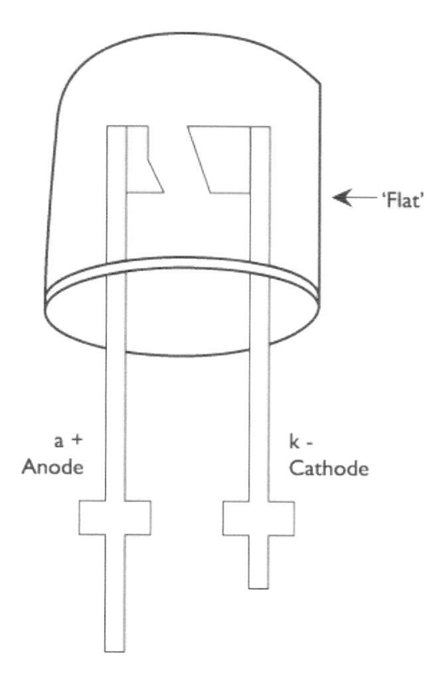


*The identifying features of a typical LED.*

As described in Chapter 4, individual decoder function outputs have power ratings in the same manner as the motor control. The exact limits will be found in the manuals of any particular decoder, which should be read carefully before starting work. In the majority of cases, the use of LEDs won't exceed any continuous rating limit, although normal lamps or bulbs might in some cases. Expect to find a limit of about 100mA and this will be part of the aggregated maximum continuous current-carrying capacity of the decoder. Note that the rated output voltage will exceed that of many LEDs and lamps, so a resistor must be included in the circuit, in series with the LED.

## LIGHTING KITS

Special lighting kits can be purchased for specific models. These have the significant advantage of a specially designed circuit board equipped with the correct resistors to suit the chosen LEDs. Such lighting kits have comprehensive instructions, not only on how to use the lighting kit and connect it to decoder, but also how to go about the practical modelling. This could be useful for modellers unsure of how to proceed with a home-made lighting project. Lighting kits usually have coloured wires soldered to the circuit board(s) to indicate which function wire from the decoder should be connected to a particular solder pad. They are generally easy to use and worth the additional cost over the purchase of individual components.

## CHOOSING AND USING LEDS

Experienced DCC users generally prefer to use LEDs when installing lighting systems. LEDs offer significant benefits over normal lamps in terms of reduced power consumption, better light emission and longevity. They have a much longer lifespan (measured in years) than conventional incandescent lamps, which may last only several thousand hours at best, and they also produce considerably less heat, which is beneficial when installing lighting systems in

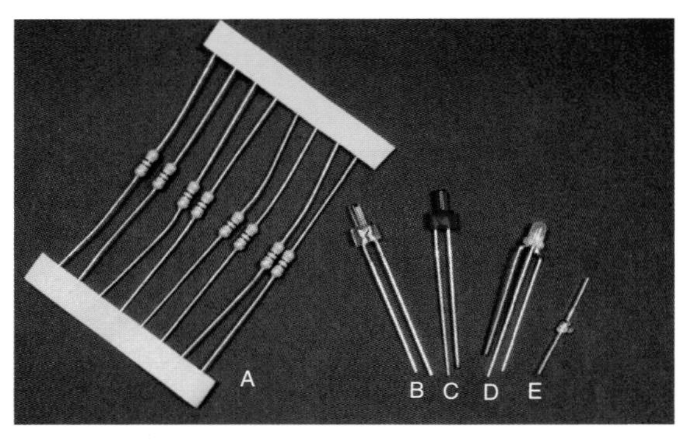


*Despite the wide variety of available LEDs, experienced modellers very quickly settle on a small number of different types – sometimes as few as three or four – and these are often used in combinations with each other and also with plastic optic-fibre light guides that are used to transmit light from the LED to the exterior of the locomotive or rolling stock bodyshell. In the main, it's the constant lighting type of LED that is used rather than those that flash or have bipolar characteristics. None the less, a project included at the end of this chapter describes the use of bipolar (bicolour) LEDs for creating the running lights in heritage diesel multiple units (DMUs). A: resistors; B: lighthouse LED; C: red LED; D: bi-polar LED; E: co-axial LED.*

plastic models. They were not always as popular as they are today: the yellow LEDs of old certainly did not look right when installed in a model and, as a result, incandescent lamps were considered to be more desirable at one time, despite their technical disadvantages.

LEDs have come into their own since the introduction of blue/white types, which look very much like the headlights of modern locomotives and multiple units, together with 'sunny white' or golden white LEDs, which better represent incandescent lights fitted to older British locomotives and rolling stock together with US-outline prototypes. Furthermore, LEDs have become much smaller in size and at a lower cost, making them much more practical to use in the smaller scales. There are lots of different types and sizes, and it takes a bit of research to decide which are likely to be useful in a particular project.

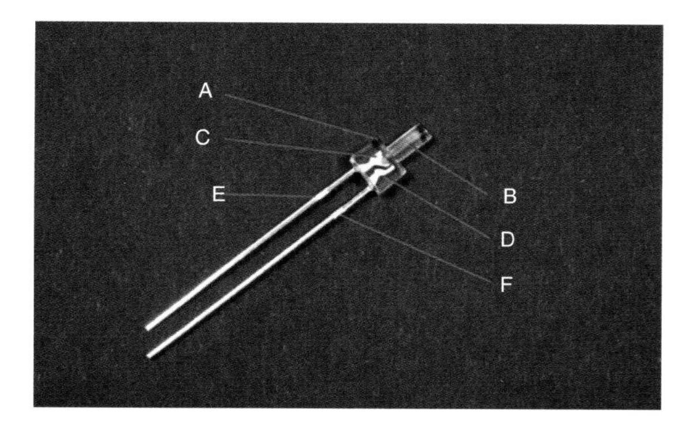


*It is very important to recognize the polarity orientation of LEDs in order to connect them the right way round to the output wires on a decoder. The long leg is usually the positive or 'anode' pole. However, that is not always the case and the best way would be to check the internal fittings, where the smaller of two metal flags indicates the anode. The anode is always connected to the blue positive common wire. (A) Body of the LED; (B) lighthouse type lens; (C) large metal flag, which is on the negative 'cathode' pole; (D) small metal flag on the positive 'anode' pole; (E) the short leg is usually the negative lead; (F) the long leg is usually the positive lead – but not always, so beware!*

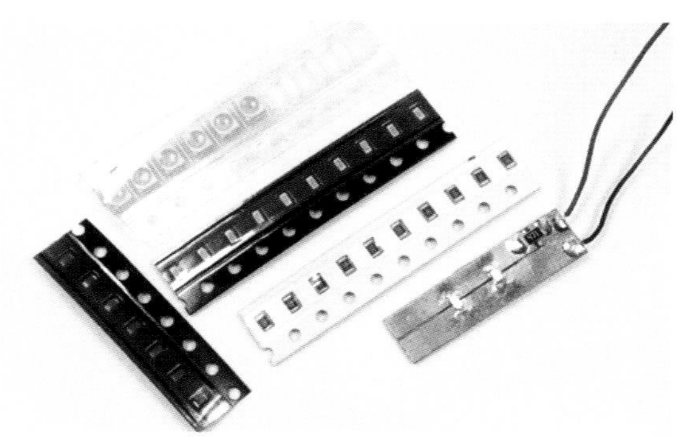

*Smaller locomotives and features such as train reporting disc headcodes require some ingenious methods to obtain good lighting effects. The tiny SMT surface-mount LED and associated resistors are very useful for installations in a tight space. Some SMT LEDs are less than 1.5mm in length and 1mm in width. They may need a heat sink to protect them from damage when soldering to a piece of PCB.*

Here is a summary of the different types of LED that have found popularity with modellers both in the UK and in the USA:

Lighthouse LED: the body of the LED is topped with a round lens that looks like a lighthouse. These find extensive use because the lens can be mounted directly in a hole drilled through the bodyshell of a model. Use those with a 2mm diameter lens. Note that they can be (carefully) filed back to reduce the diameter of the lens.

Coaxial LED: these are similar to the lighthouse LED, except the lens is usually shortened and domed in shape. They are not quite as useful as the lighthouse type.

SMT LED: tiny surface-mount LEDs are now readily available to railway modellers and come in a variety of colours including blue/white and natural 'sunny white' or golden white to better represent incandescent lighting.

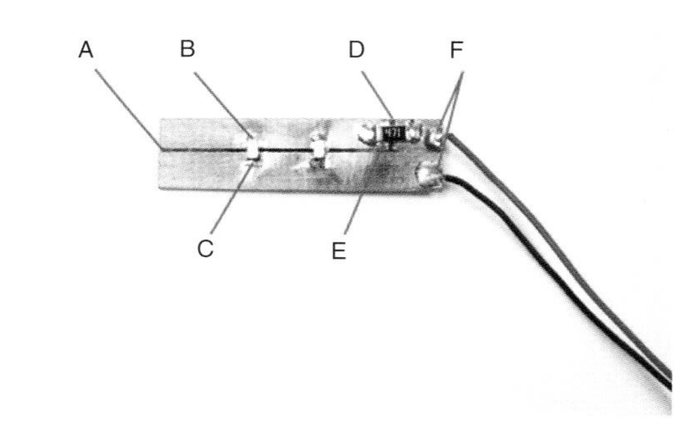


*This picture shows two SMT white LEDs soldered to copper laminate circuit board. It is the most reliable way of fitting them to a model, although it is possible to solder leads to the LED terminals (B and C), if you are very careful. The LEDs are soldered across an insulation gap cut in the circuit board (A) and the same is done to the tiny chip resistor (E), as shown at (D). The leads are soldered to the edge of the circuit board away from the LEDs (F). When using SMT LEDs, it is important that you check which pole is positive (the anode). They vary from manufacturer to manufacturer, although the ones I choose to use have a small green triangle on the rear that indicates the positive pole.*

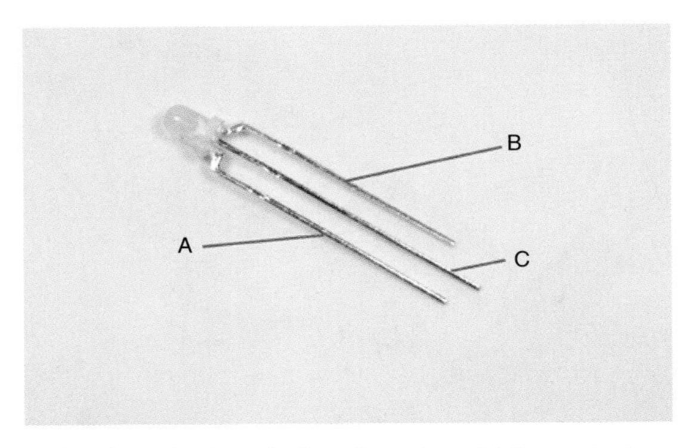


*A bicolour LED with three legs: the middle one is the positive common (C), while the other two operate one or the other colour light emissions from the LED and are the negative leads (A and B). Some bicolour LEDs that appear to be identical to that shown here have two positive leads and a common negative lead, which is of little help to DCC users because the wiring harness on decoders is the opposite. Ensure that you buy bicolour LEDs with a positive common that can be connected to the blue wire.*

They do not have 'legs', but are designed for soldering to a piece of circuit board. You have to be extremely enthusiastic to use them because they are tiny and difficult to handle. None the less, they find applications in locomotives with minimal space. I have used them to install lighting in some incredibly difficult spots in small locomotives.

Bicolour LED: depending on which way the current flows through the LED, these display one of two colours, usually yellow and red.

## TOOLS AND MATERIALS TO HAVE AT HAND

Digitally controlled lighting projects, whether they are hard wired or assembled with the use of a kit, are more involved than plugging a decoder into a DCC-ready model with a NEM 652 interface socket. You should have the following to hand, many of which should be among the basic supplies generally found on any modellers' workbench.

- Electrical insulation tape.
- Self-adhesive sticky pads.
- A selection of fine, flexible 0.6mm electrical cable.
- Small diameter heat-shrink sleeve.
- Miniature plug and sockets for multiple units.
- Phosphor bronze strip for additional pickups.
- PCB strips (i.e. copper-laminated sleeper strip) and copper laminate sheet.
- Soldering iron and electrical (lead-free) solder/flux.
- Wire strippers.
- Adhesives.
- Jeweller's screwdrivers.
- Assorted pieces of scrap black styrene.
- Tweezers, wire cutters, fine-nosed pliers, wire strippers and a selection of files.
- Black bathroom sealant.
- Modelling scalpel or knife.
- Piece of junior hacksaw with the end covered with tape to protect the hands.

## BASIC DECODER TO LIGHTING CIRCUIT WIRING

Theoretically, wiring up lighting to decoders is very simple because most of the electronic circuitry designed to switch the functions, and therefore the lighting effects on and off, is contained within the decoder components. One of the problems the modeller has to overcome is that the function output voltage will be at the same voltage destined for the motor (12V dc), which is considerably more than that required by an LED or incandescent bulb, such as a grain of wheat bulb. A resistor is built into the circuit to reduce the voltage and protect them from burnout. The placement of resistors is quite important, depending on the type of lighting circuit you are constructing.

While the orientation of incandescent light bulbs is unimportant, provided they are protected by a resistor, the situation is entirely different when using LEDs, which are polarity dependent with positive and negative poles. Such dependence must be observed when making up lighting circuits. This relates to the

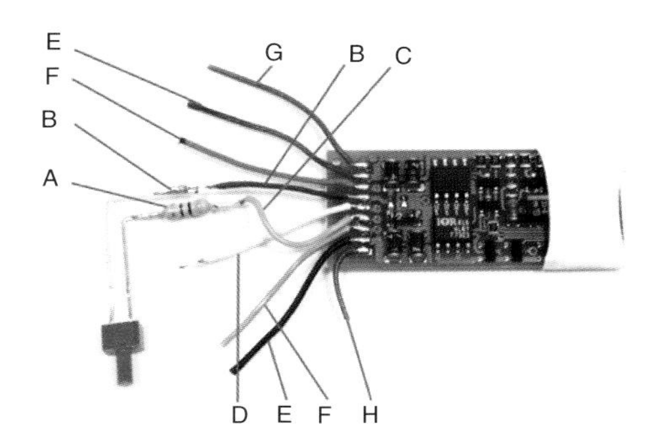


*Wiring lighting circuits to a decoder is, in principle, very simple, as shown here. Projects only become more complex as further LEDs are added for twin lights and other features. The negative pole of the LED is connected to the negative return cable for function output F0R (C) with a resistor in series to protect the LED (A). The positive pole of the LED is directly connected to the blue positive common wire (B) of the decoder (it is a common to all of the function wires). The other coloured wires are: (D) function F0F or output A for headlights; (E) wires to be connected to the track supply are red and black; (F) wires to be connected to the motor terminals are orange and grey; (G) function 1 or output C is the green wire; (H) function 2 or output D is the purple wire.*

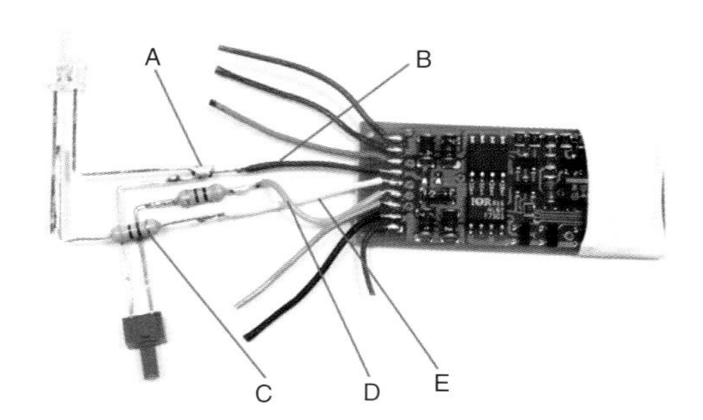


*Here is the same decoder fitted with a headlight 'white' LED connected to the white output wire and tail-light red LED connected to the yellow function wire. Note the common positive connection between both LEDs (A) and the blue wire (B). Resistors are used in series in the circuit, ideally one per LED, for maximum protection (C). The function output wires (D and E) are in fact negative returns. Note that one is 'forward' lighting and one is 'rearward' lighting, so the front and back of every model has to be determined if they are switched with the reversing switch on the throttle and turned on and off with the F0 button.*

decoder function wires, which share a common positive wire to complete the circuit. As a result, the anode of the LED must be connected to the positive common wire and the cathode connected to the negative return wire appropriate to the function, otherwise the LED will not be illuminated when the function button is pressed. At one time I preferred to wire my resistors in series with the blue positive common wire on the decoder. I very quickly realized that it was more desirable to wire the resistor in series with the negative function wire, providing one resistor for each LED.

## LIGHT LEAKAGE

One of the irritating practical problems with lighting systems is 'light leakage'. LEDs can be very bright and light produced in the body of the LED can leak, even though the LED lens will capture and direct most of the light in the direction it is facing. Although LEDs are far better than incandescent light bulbs for directing light in a beam, there is always some light that escapes into the cab or car body of the locomotive. That light can also appear as a glow through lightly coloured plastic bodyshells such as those once offered by Lima and through the cab windows of any model in a most untypical way unless something is done to seal it in.

There are two solutions to this problem. Firstly, after preparing the bodyshell to accept whichever lighting method you have chosen, paint the inside surfaces black. You may also apply thin silver foil. Once the LEDs or bulbs are installed, paint them so that light does not leak away. The second option is to create a physical barrier with insulation tape (not always that reliable) or other sticky material such as black sealant.

*Black bathroom sealant is an excellent product for preventing light leakage into the cab or bodyshell of a locomotive. It can also be used to secure LEDs in place without having to use glue.*

The problem with paint is that it can migrate through the holes in which bulbs and LEDs are fitted, marring the exterior finish of the model. That is why I prefer to use flexible plastic plumber's sealant, which is soft, pliable and easily worked around the LEDs. It has been a far more reliable method because it also acts as an adhesive, is temporary and can be removed without leaving a mess behind. A typical brand of bathroom sealant is Homelux, available in DIY stores.

## INSTALLING LIGHTING IN ROLLING STOCK

You can control lighting in rolling stock using a decoder and its functions in the same way that you would in a locomotive. This brings the opportunity to control interior lighting, door lock lights, tail-lights and, in the case of research and departmental vehicles, the lighting that may be used for lineside, track and tunnel inspection. In fact, if the budget is not an issue, you can do an awful lot to create a very interesting collection of rolling stock with fully controllable lighting simply by using decoders.

UK-outline modellers often restrict decoder-controlled lighting in rolling stock to tail-lights, so they can be switched on and off together with a representation of portable flashing tail-lights by using the strobe feature now found on many decoders. The methods of installing these lighting effects are exactly the same as working on locomotives: you will need to protect the LEDs with a resistor, which will be wired in series with the function output wire, and a reliable connection from the track to the interior of the vehicle will have to be established through the wheels to ensure that power to the decoder is not interrupted, which would result in the lights flickering. In some instances, modellers may decide that switched interior lighting is not absolutely necessary, avoiding the use of a decoder by using a bridge rectifier instead, from which a suitable power supply is tapped. A word of warning, however, on the use of bridge rectifiers in plastic models. They can generate heat – sometimes a great deal of heat: even with the use of a heat sink they are capable of damaging rolling stock, so be careful how you combine certain electrical components in plastic models. Assemble and test circuits on the workbench first.

## ADDITIONAL PICKUPS

Under the heading of 'additional pickups' we should also place 'suitable wheels'. This is not as daft as it sounds, simply because many rolling stock projects will require suitable wheels that are capable of conducting current from the rails to pickups. It is generally accepted that most locomotive models will have sufficient pickups and appropriate wheels because they will be necessary to operate the model on analogue control anyway. When it comes to trailer cars on multiple units and rolling stock not equipped with factory-installed lighting, such as that fitted to Bachmann Pullman coaches, manufacturers do not see the need to provide current pickups or wheels to work with pickups. You have to equip your rolling stock with those instead.

It is possible to add simple pickups by soldering pieces of phosphor bronze, which is usually quite springy, to copper-laminated strip, which also works as a convenient solder pad to connect the wires that lead to the decoder

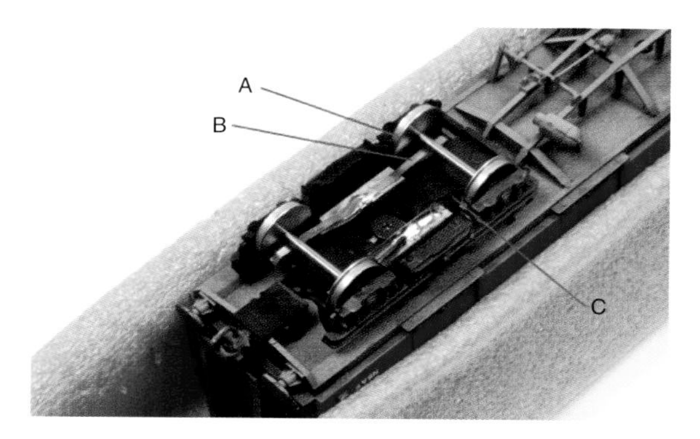


*Simple current pickups can be fitted to most rolling stock bogies without having to make complex modifications. This Bachmann Super BG model is fitted with working tail-lights. Turned metal wheels with a large contact area on the rear of the disc (A) are used. One side is live, including the axle, which means one pickup can be made to work on the axle, reducing drag (B). The opposite wheel is insulated from the axle and the phosphor bronze pickup wiper acts on the back of the wheel (C).*

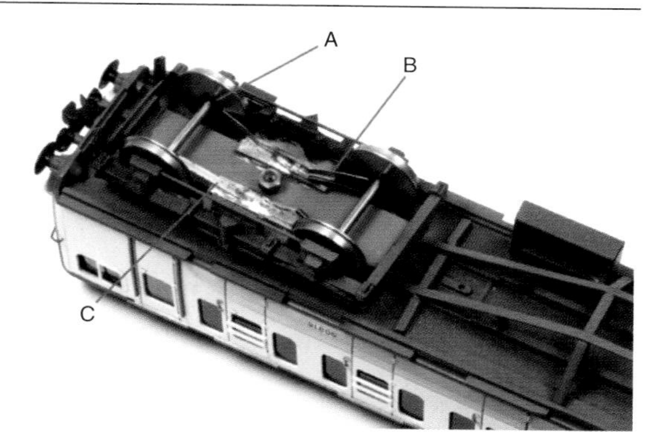


*Multiple unit trailers fitted with decoders and lights should be treated the same way as unpowered rolling stock. The insulating bush that isolates the wheel on one side from the live axle is seen at (A). Phosphor bronze pickups act on the large contact area of the wheel disc (B). The other pickup acts on the axle, which is electrically live with the opposite wheel (C).*

within the coach or wagon body. The wheels themselves must have a reasonable area on the rear face against which the phosphor bronze pickup can act, so that current can be reliably collected from the rails and conducted to the pickups and internal wiring. Whilst this might seem to be a simple point, there are frequently problems with such pickups and it's worth taking time to ensure that they act correctly against the back of the wheel and there are no burrs where the phosphor bronze has been trimmed before installation, which could do much to interrupt the power supply.

## DITCH LIGHTS ON US LOCOMOTIVES

Ditch lights are two running lights fitted at one or both ends of a locomotive (depending on the locomotive type – use reference photographs) to supplement the main headlights, although their function is different to the headlights. The light beams are angled so they cross and have a different light intensity when compared to the headlight(s). Ditch lights were introduced as part of grade (level) crossing safety measures to help motorists to better determine the location and speed of approaching trains. They are constantly lit until the horn and bell are activated, when the lights then flash in an alternating pattern – left, right, left, right or together, depending on the practice of individual railroads. Not all railroads, however, have a flashing option with their ditch lights when the horn is sounded and the modeller should research the operating rules of the chosen railroad carefully when fitting ditch lights and programming decoders to make them work.

Wiring ditch lights to a decoder to achieve the optimum lighting effects is an interesting task. Most decoders offer a setting in their auxiliary functions for alternately flashing lights, which would appear to suit ditch light operation quite well. Unfortunately most HO/OO 1 amp decoders have only four output terminals suitable for lighting functions and examples of GP38-2s I have modelled (amongst other locomotives) are equipped with ditch lights at each end. Of those four function outputs, two are required to control the locomotive headlights.

This is when reading decoder manuals can be very useful in determining what can be achieved. To have flashing ditch lights on either end of a general purpose locomotive, together with operational headlights, would require a six-function decoder such as the Digitrax DH163, or two separate decoders, one of which would be reprogrammed to function map the output functions to different buttons on the throttle.

It is worth noting that large road diesels have ditch lights only on the front. As a result you will get away with four-function decoders, assuming that they have to be made to flash at the press of a function button. For example, one light would be wired to output C, set for flashing mode. The other would be wired to output D and set for inverse flashing, so they appear to be working alternately. The exact details would have to be gleaned from the decoder manual, especially with regard to choosing the right CVs to change to achieve the desired effect. In N gauge, modellers may find the restricted number of decoder functions on the typical micro decoder will determine if flashing ditch lights can be incorporated. Most N gauge models of US-outline locomotives rely on one LED to light both the headlights and ditch lights, so there are practical difficulties to overcome, space being the least of them!

## FUNCTION MAPPING: MAKING THE LIGHTS WORK INDEPENDENTLY

Programming is one of the things about DCC that cause concern for some modellers. It is perhaps unfortunate that the word 'programming' has been used, because you are not really programming the decoder in the strictest sense of the word: you are changing settings within the programming of the decoder. Function mapping is part of that process and it is important because, to make each lighting feature work independently at the press of a button on the throttle, it is necessary to allocate each function output to a function button.

All decoders come with default settings for function output to function button control. For example, a typical four-function HO/OO gauge mobile decoder has default settings that control function output A and B with button F0 and the direction change (reversing) switch. Outputs C and D are controlled by buttons F1 and F2. It may be desirable on a Lenz LH100 throttle, for example, to control outputs A and B with buttons F1 and F2, with the remaining two functions controlled by buttons F3 and F4.

Some lighting projects require two decoders with a shared address and care has to be taken when programming them, for example to change the values in the appropriate 'configuration variables' (or CVs for short). The mobile decoder responsible for controlling the motor in a model should be considered the 'primary' decoder and it is usually installed straight into the model in its default form, connected to the motor and tested on the service track.

The 'secondary' decoder is usually a mobile decoder, sometimes older and cascaded from front-line use with its motor functions unused or a specific function-only decoder acquired specifically for the project. This decoder has to be programmed prior to installation or the programming instructions sent to it, if installed in the model alongside the primary decoder, will also be accepted by the primary decoder. To do that, its motor output wires are connected to a spare electric motor or a resistor and light bulb to provide an electrical load that the command station would recognize during programming. The decoder can then be connected to the programming wires of the command station and tested in the normal way. If there is no electrical load for the command station to detect, it may assume that the decoder has been 'installed incorrectly' and will show an error message, and the secondary decoder may not accept programming commands in the process.

The secondary decoder has to be programmed to change the function outputs for the yellow and white wires (outputs A and B). This process is called 'function mapping' and it simply allocates the output(s) of the decoder to different function buttons on the hand throttle so that they do not conflict with those used by the primary decoder when both decoders share the same address.

The primary decoder is also subject to function mapping to allocate function outputs A and B to function buttons 1 and 2 (yellow and white wires). This means that the decoder default setting is changed and those outputs are no longer controlled by a change in the direction of travel or switched on and off by function button 0. This seemingly complex process is not as difficult as it might appear and the instructions supplied with modern decoders will give details of which CVs to change and what values are required for function mapping.

When two decoders are installed in one model, they will both take up the same programming when the model is placed and programmed on the service track. This is why it is important that some programming of the secondary decoder is done before it is installed in the model, either by connection with the service-track wires or in a special programming device that can be connected to a personal computer. Some programming of the secondary decoder could be done as 'programming on the main' (PoM) if it has an address that is unique to itself, that is something other than a default address such as 0003. Once both decoders have been function mapped to enable independent control of all six lighting functions, they can be programmed with a shared address so that they can be controlled together.

## US DIESEL LOCOMOTIVE RUNNING LIGHTS AND DITCH LIGHTS (HO SCALE GP38-2 LOCOMOTIVES)

One of the biggest challenges of this project is finding a simple but effective way of installing ditch lights. On Soo Line GP38-2s, the chosen project models, they are mounted on the top of the anti-climber at each end of the locomotive, which presents some challenges with hiding the wires and gluing the light casings securely in place. Square-mount LEDs were used instead of white metal castings for the casings on the Athearn model and the wires passed through a hole in the pilot immediately below the anti-climber. The Proto-2000 was equipped with white metal cast ditch lights, which proved particularly challenging to fit with SMT LEDs. The practical techniques used to fit lighting to these two models is not exclusive to them and they can be adapted to suit other locomotives.

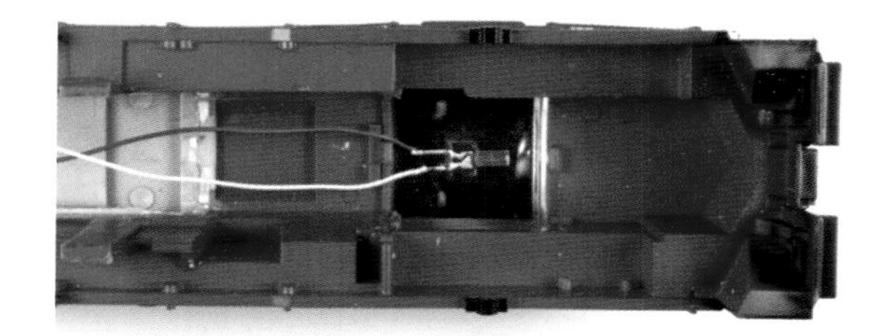

*After the Athearn model was stripped of its rudimentary lamps and lighting circuits, it was equipped with 'sunny white' LEDs, which were used to light the headlight/rear-light lenses at both ends of the model. This picture shows the LED fitted within the cab. Black insulation tape is applied to the inner surfaces of the model to prevent light glowing through the plastic.*

*One of the biggest challenges of this project is finding a simple but effective way of installing ditch lights. On Soo Line GP38-2s, in common with many other railroads' locomotives, they are mounted on the top of the anti-climber, which presents some challenges with hiding the wires and gluing the light casings securely in place. One option for creating ditch lights is to use square-mount LEDs instead of white metal castings for the casings and pass the wires through a hole in the pilot immediately below the anti-climber.*

*The resulting appearance is quite neat. When the model is detailed, the square-mount LEDs will be enhanced with a representation of the light cover and the support bracket on the underside of the anti-climber. A dab of paint will complete the picture.*

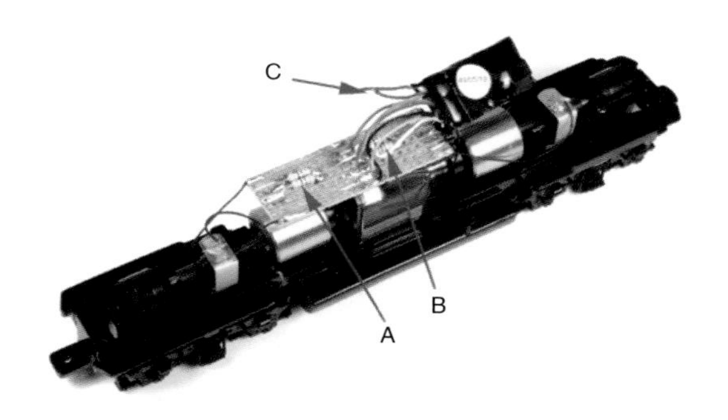


*The design of the Athearn chassis does not provide any secure surface to mount the decoder. This picture shows the completed arrangement with circuit board providing connections for the decoder harness and connections for the various lighting circuits. The resistors for each lighting circuit can be seen at (A) and (B). Part of the project was to fit digital sound, and the separate leads for the speaker are visible at (C).*

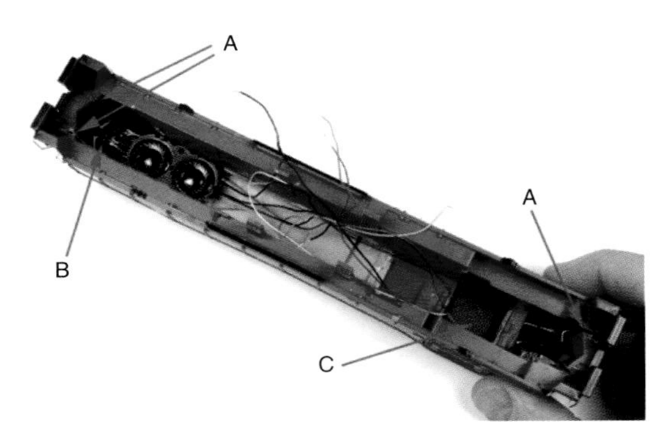


*Internally the bodyshell wiring looks like this. The wires leading to the ditch lights pass through the pilots at (A) to emerge just below the front and rear anti-climbers. The long bonnet headlight/number board light is located at (B) and illuminates the number boards as well. The cab headlight is hidden behind a piece of black styrene at (C).*

*This picture shows the arrangement in a Proto-2000 GP38-2 model, which is similar to the Athearn model. The headlights (A) and adjacent number boards are illuminated with the 'sunny white' LEDs. Note the ditch lights with leads fed through the front pilot as discretely as possible (B). They were concealed behind casing brackets when the model was finally detailed.*

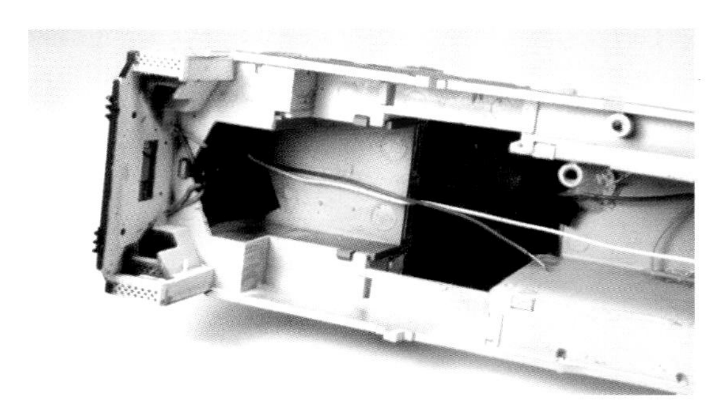

*The ditch light leads are fed through the pilot into the interior of the Proto-2000 bodyshell. Because the Soo Line unit I chose to model for this particular project does not have flashing ditch lights, both LEDs were connected together and powered by one decoder function output.*

## BRITISH STEAM LOCOMOTIVE LIGHTS (HORNBY A4)

Recreating the warm light from oil lamps is particularly challenging when you consider that we are proposing to use modern technology such as light emitting diodes (LED). With a little bit of ingenuity, it is possible to create an effective-looking oil lamp from the body and legs of an LED if the right colour is chosen. Some after-sales detailing manufacturers, particularly those specializing in lighting kits, offer their own version of an oil lamp that can be built into the front of a steam locomotive model relatively easily. All that remains for the modeller to do is to work out how to route the wires from the oil lamps back to the main circuit board of the locomotive.

This was part of a larger project to install lighting and steam effects in a Hornby A4 Pacific locomotive. While the section that deals with the installation of a steam/smoke generator is covered later (*see* Chapter 6), the fitting of oil lamps is covered here.

One of the features of oil lamps is that they are not turned off with a flick of a switch like an electric lamp. This creates an interesting detailing feature that DCC enthusiasts must take into consideration, for when a steam locomotive is reversed, say when shunting a rake of coaching stock into a siding or platform, the oil lamps do not extinguish as the reversing lever is changed. They remain continually illuminated until they run out of oil! When programming the decoder after the project has been completed, the modeller will have to consider function mapping outputs A and B from the decoder so that they are illuminated regardless of the position of the reversing switch on the hand throttle or controller.

To achieve the desired effect, the decoder was reprogrammed so that the oil lamp LEDs would remain illuminated in spite of the direction of travel, yet still switched on and off by function button F0. Normally, F0 forward is controlled in CV33 in a Lenz Gold decoder (the chosen decoder for this project) and that is set to a default value of 8, which means that it controls function output A. That output is connected to the white wire on the decoder harness. When the direction change switch is thrown to reverse the locomotive, it activates function F0 reverse, which switches output B (yellow harness wire) on and output A is switched off. F0 reverse is controlled in CV34, which has a default setting of 16. Change that setting to the value of 8, which is the same as CV33, and the oil lamps will remain illuminated regardless of the direction of travel, thus solving the problem and avoiding embarrassment.

The following illustrations demonstrate how simple oil lamps constructed from LEDs were installed.

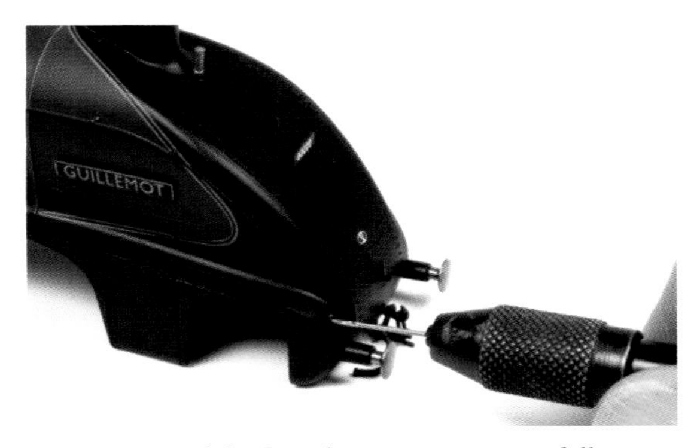


*The position of the front lamp irons was carefully determined by reference to photographs of the prototype locomotives, and a small hole of 0.8mm was drilled through the bodyshell.*

*The hole was carefully widened so that it would accept two leads from the LED 'oil lamp'.*

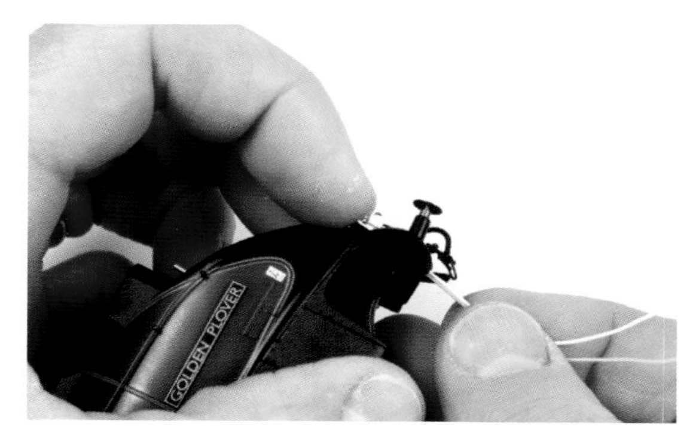


*Once in place, a small spot of glue was used to secure leads from the inside of the locomotive body.*

*A small amount of black paint was used to touch in and disguise the legs of the LED, which also acts as the lamp iron detail. The electrical fittings actually look quite inconspicuous in this simple lighting installation.*

*The result is a soft amber glow from each LED, which when painted white also offers a good representation of locomotive oil lamps. This has done much to improve this model's appearance. Judging by the position of the lamps, it has been prepared for use on express passenger and mail trains.*

## BRITISH DIESELS WITH MARKER LIGHTS (BACHMANN WARSHIP)

Marker lights applied to older British diesel-electric and diesel-hydraulic locomotives represent something of a challenge to the DCC enthusiast, because the marker lights on the full-size locomotives are quite small. Locating LEDs that would be suitable for this sort of project is not that straightforward. Consequently a number of other techniques have to be used: one such is to 'pipe the light' from LEDs located within the bodyshell to the exterior of the model.

Locomotives that typically have this marker light arrangement include some Class 20s and Class 40s, together with 'Claytons', which have marker lights combined with train headcode reporting discs. Class 08/09 shunters can be included in the list and a small shunter with accurately modelled marker lights is impressive.

Many locomotives with headcode discs may have a single white and a single red marker light located at each end. This includes our project locomotive, the Bachmann Warship diesel-hydraulic locomotive. The following project shows a useful technique that can be used to illuminate small marker lights, which would include the red tail-lights on Class 25, 37 and 47 locomotives, for example. The principle described here, which is a combination of plastic optic-fibre light guides and SMT LEDs assembled to a mini-light board, could be used to fit marker lights to any British outline locomotive. Here goes!

*Western Region diesel-hydraulic locomotive No. 818 is represented by the Bachmann 00 gauge model, which has been in production for some considerable time, albeit with a number of improvements and upgrades to improve its performance. The fitting of realistic-looking marker lights to help bring it further to life is a worthy subject for our attention.*

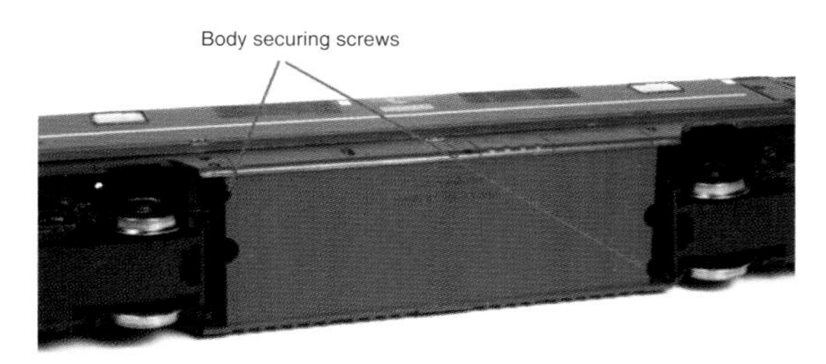


*The locomotive body is secured with two screws, as indicated by the red lines. Do not release the screws located centrally on the underframe, which only release the fuel tanks.*

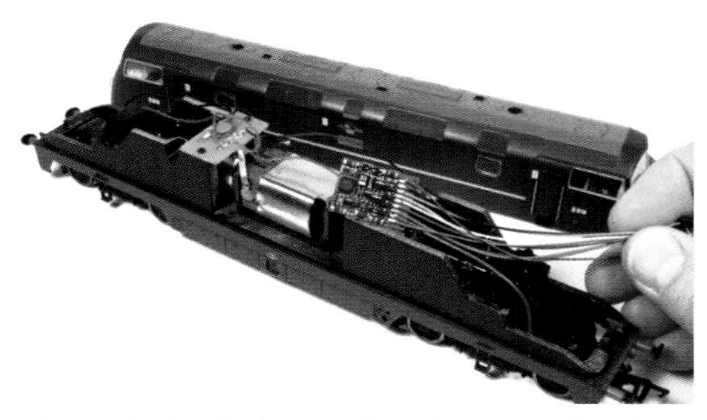

*This model has the luxury of ample space within the bodyshell and there is plenty of room to fit a high-quality, high-specification decoder.*

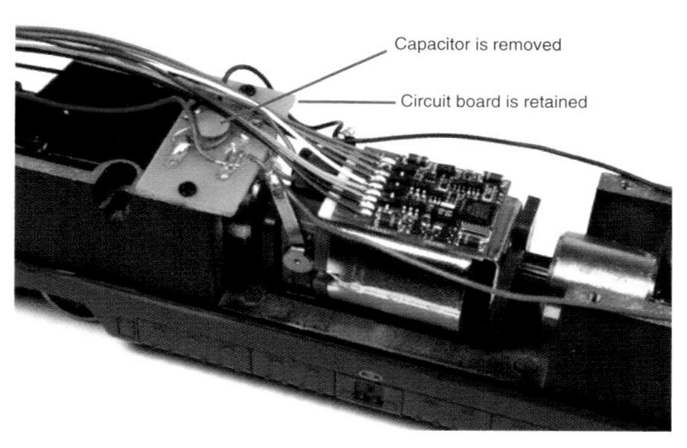


*An immediately logical place to locate the decoder is on top of the flat can motor. This photograph shows the decoder being tested to see how it fits. It is important to note, however, that the metal casing of the motor could damage the decoder if it is not covered with insulation tape.*

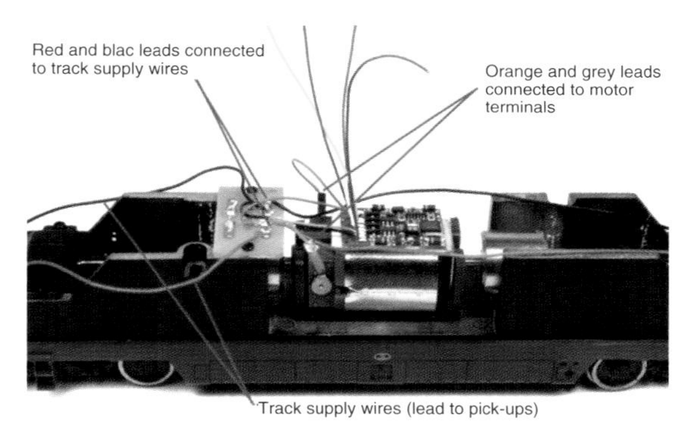


*To make life easier, always try to use the interior fittings provided by the model manufacturer. The small circuit board provided in the model was retained because it provided a good place to connect the red and the black wires from the decoder to track supply pickups. This saved time and some fine equipment wire. The orange and grey leads were connected to the motor terminals and the model tested on the service track to ensure that the decoder installation was done correctly. What should also be tested at this stage is that the locomotive runs in the correct direction. For example, when the hand throttle is set to the forward direction, the locomotive should run in the same direction as the rest of your collection. This can be checked using a 'control' locomotive fitted with a DCC interface socket.*

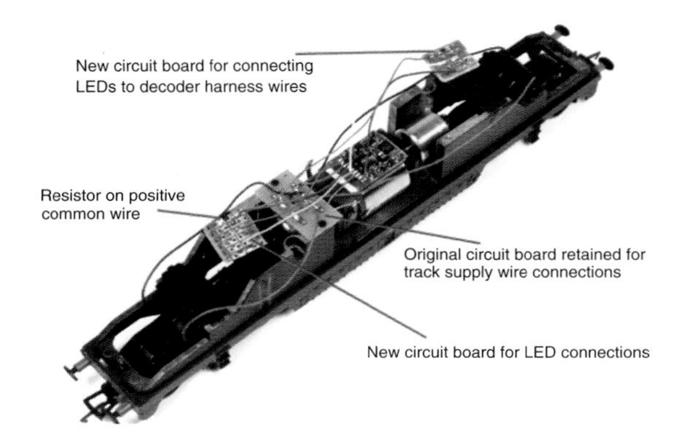


*Two small circuit boards were prepared as secure points for connecting the wires from the LEDs and connecting the resistors within the electrical circuits, as shown here. These were connected to the decoder function output wires and positive common blue wire.*

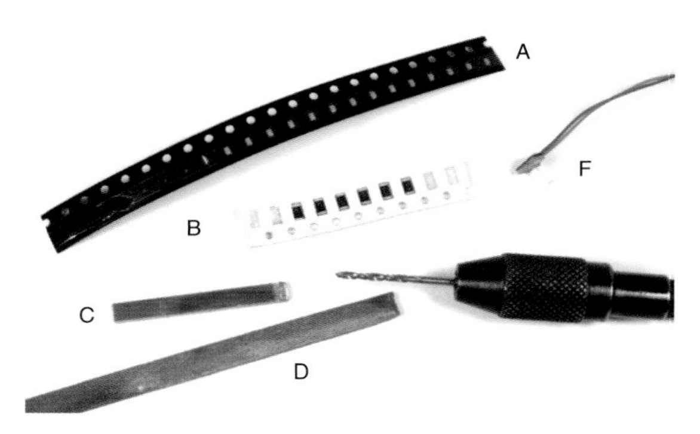


*The components for assembling the LEDs and light guides to create mini light boards. White and red SMT LEDs (A) were used as the light source, with micro-resistors (B) on the circuit boards to save space. Copper laminated strip was used as suitable material upon which the mini light boards were constructed (C). Essential tools include a fine cut file (D), high-speed twist drills and a pin vice (E). The completed light source assembly is shown to the right (F). The following photographs demonstrate how the light sources were constructed.*

*File an insulation groove into the surface of the copper laminated strip to isolate the poles of the SMT LEDs from each other.*

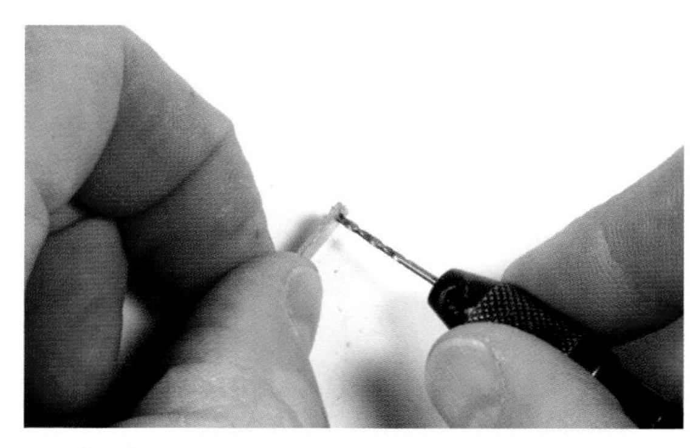

*Drill a hole 0.8mm in diameter through the middle of the groove previously filed in the copper laminated strip.*

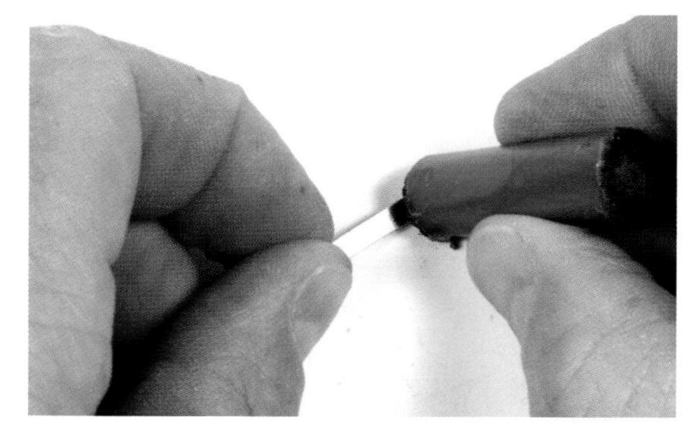

*A fibreglass pencil or rubber is used to burnish the copper laminate to ensure that the leads and LED will solder securely to it without the risk of a dry joint.*

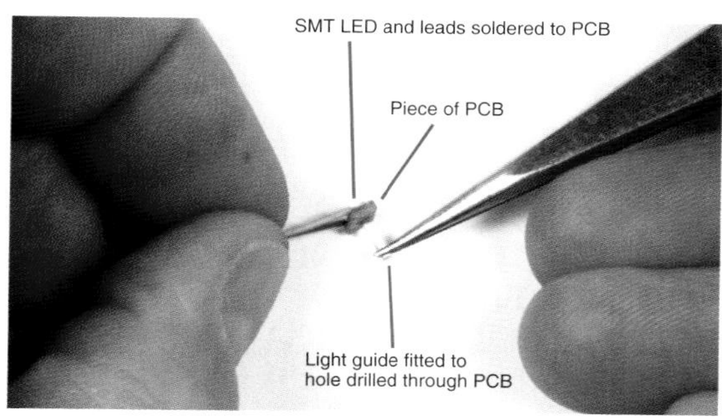


*Assembly of the mini light boards is completed by soldering the SMT LED across the isolating groove, soldering the leads to either side of the LED and inserting a small piece of optic-fibre light guide from the rear of the circuit board. This is secured with superglue.*

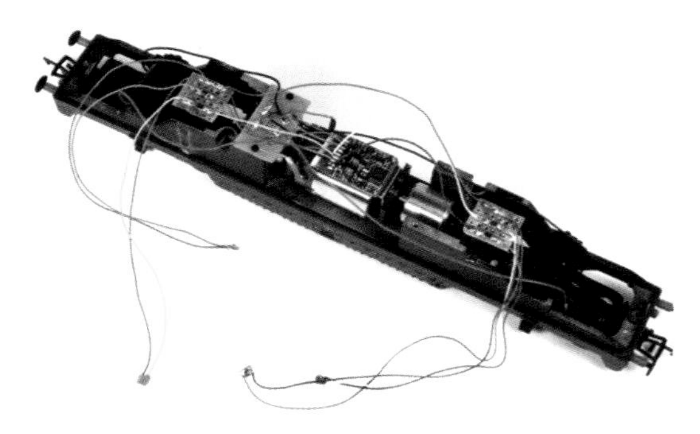

*Each light board is provided with sufficient lengths of wire to reach small circuit boards that were constructed earlier. The circuit is completed by soldering the ends of the leads to the circuit board, noting which lead should be soldered to the positive common terminals.*

*A quick test in low lighting conditions shows how the light is transmitted through the light guide and therefore through the bodyshell of the locomotive – or so one hopes!*

*We now turn to modifying the locomotive to accept the recently constructed light boards.*

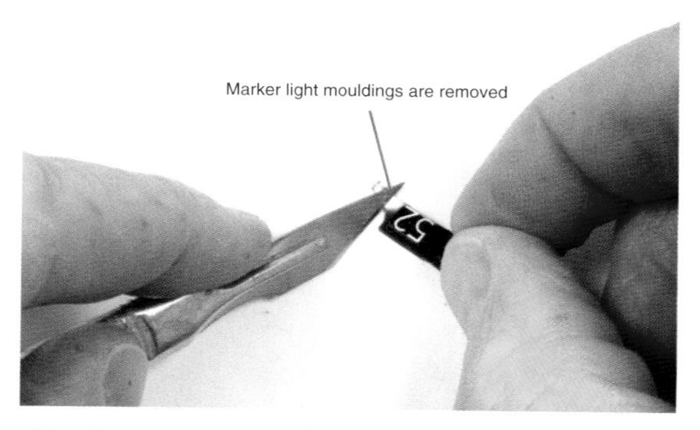


*The dummy marker light lenses provided with the locomotive are moulded as part of the headcode panel. These should be removed with a sharp scalpel to make room for the new light boards.*

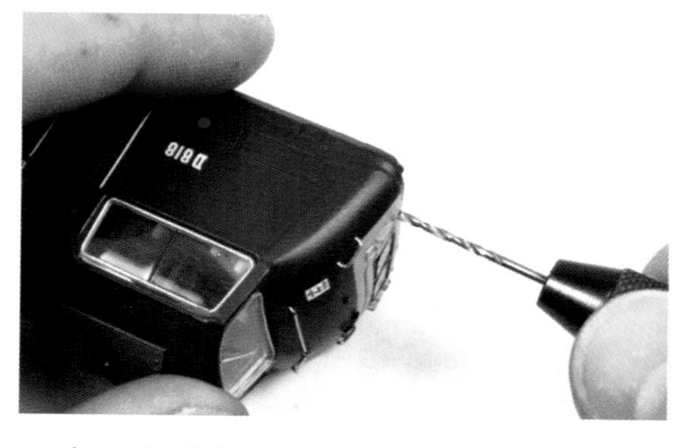

*Each marker light is opened up with the same drill bit that was used to drill a hole through the copper laminated strip in preparation for installing the light guides.*

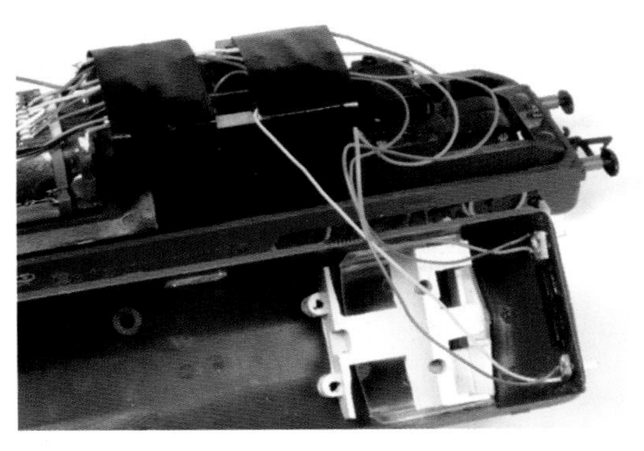

*Each light board simply plugs in from the rear, with the light guide protruding through and beyond the bodyshell.*

*This image shows the completed project being tested before the light guides are trimmed back to the surface of the bodyshell with a sharp scalpel. The light is transmitted very effectively and the design of the four light boards ensures that there is minimal impact on the detailed cab interior of the model.*

To finish the project, the locomotive could be programmed so that the function outputs are function mapped to enable the independent control of all four marker lights. This would require the reallocation of outputs C and D to function buttons F3 and F4 on the controller, which would still allow independent control of the red tail marker lights. That releases function buttons F1 and F2 for the control of the forward-facing white marker lights. Note that the value of zero would have to be programmed into the function mapping area, which would normally permit control of function outputs A and B with button F0 and the direction change switch so there is no conflict with buttons F1 and F2.

## BRITISH DIESELS AND ELECTRIC LOCOMOTIVES WITH HEADCODE BOX AND DISC INDICATORS (BACHMANN CLASS 20)

The techniques used to install marker lights in the Bachmann Warship model (*see* above) can be applied to a number of other OO gauge models. The English Electric Type 1 Class 20 is available as a OO gauge model by Bachmann and is certainly one of the more popular models in the range. In 2006/7 it was released in several versions with digital sound, but at no time has it been equipped with lighting circuits. Class 20s were built with two types of train reporting: discs with marker lights and a train reporting code headcode box. Latterly, rebuilt locomotives in use with Direct Rail Services (DRS) have used modern light clusters to ensure they comply with contemporary standards, which is not covered in this project. Some Class 20s received high-intensity headlights in a square housing and the position of these varied somewhat from locomotive to locomotive.

The Class 20 is a prime example in which lighting boards equipped with tiny SMT LEDs could be used to minimize the amount of space used by the lighting circuits and to reduce the impact in the cab area of the model, where the detail is considered to be particularly good. As mentioned above, many of the Class 20 locomotives built by English Electric in the 1950s were equipped with headcode indicator discs, which can be folded up to reveal a marker light. Furthermore, the tail-lights in both this version of the Class 20 and the type fitted with headcode boxes are small marker lights, which are certainly too small to be represented by lighthouse LEDs, even if the lens were to be carefully filed down. The techniques described above could be used to create a home-made lighting project for the Bachmann Class 20.

Alternatively, given the tight confines in the model, lighting kits may be used instead. The one demonstrated with this project was obtained from Express Models, a small after-sales detailing manufacturer that offers kits for both the headcode disc and headbox versions. Both kits rely on circuit boards fitted with SMT surface-mount LEDs (called light boards) that squeeze neatly into the restricted space at the bonnet end and have minimal impact on the cab interior. They hold sufficient LEDs to illuminate all of the forward disc marker lights and both tail-lights – using all four functions on a decoder to operate them Consequently, all of the forward marker lights are illuminated and not individually switched with this kit, leaving it to the modeller to decide, at the time of installation, which of the marker lights will be illuminated (usually the lower two on the outside). Those that are not required should be covered with a small piece of black insulation tape or you should avoid drilling out those marker lights were lighting is not required. The kit does not include a circuit for a high-intensity headlight, which would be required for some members of the class in their later years.

The following sequence demonstrates how both versions of the model may be fitted with a decoder and bespoke lighting kit. The model can be enhanced with etched headcode disc details instead of using the bulkier plastic versions supplied by Bachmann.

*Class 20s have certainly made a comeback in recent years. In 2007 many were either in private ownership or located at preservation sites, such as Barrow Hill in Derbyshire. You are looking at a Class 20 locomotive equipped with a full set of headcode discs.*

*Headcode disc detail in close-up showing the marker light located behind the folding metal indicator disc.*

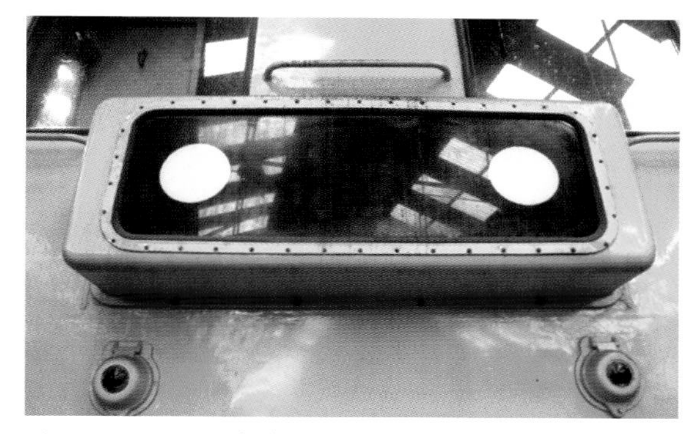

*As construction of Class 20s progressed, the indicator discs were replaced with headcode boxes. The two tail-lights are still of the marker light type.*

*A pair of blue TOPS (Total Operations Processing System) Class 20s is a good example of typical motive power used on 1980s MGR coal trains and freight trains. Overall the character of this class has been captured by the Bachmann model despite the lack of lights, which provides the opportunity for an interesting DCC project. To run both these models as a matched pair with DCC consisting and working lights will recreate a classic motive power combination that was popular with most enthusiasts.*

*To commence work, you need to have free access to the interior of the model. Screws are used to secure the body to the chassis rather than bodyshell clips. The Class 20 has two retaining screws, one located at each end of the model.*

*Removal of the cab is straightforward, despite the close-fitting nature of the moulding. Two clips at the base of the cab secure it to the bonnet. Gently tease the cab sides away from the bonnet until the retaining clips are clear of the sockets in the bonnet. Gently pull the cab off and place to one side.*

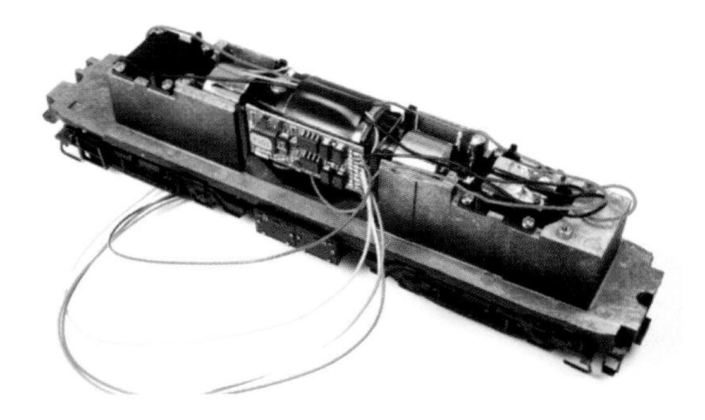

*Since space inside the long narrow bonnet is so tight, a single-sided decoder was chosen for this project. This is installed first, and then insulation tape is applied to the side of the motor after a small amount of plastic has been trimmed from the motor cradle. The decoder is fitted with double-sided adhesive tape and not a sticky foam pad. Note that the four main decoder wires are soldered to various terminals. The red and black wires go to the track supply via a simple circuit board made of PCB strip. Orange and grey wires go the other way, hardwired to the motor terminals. At this stage it is wise to refit the body temporarily to test the fit of the decoder.*

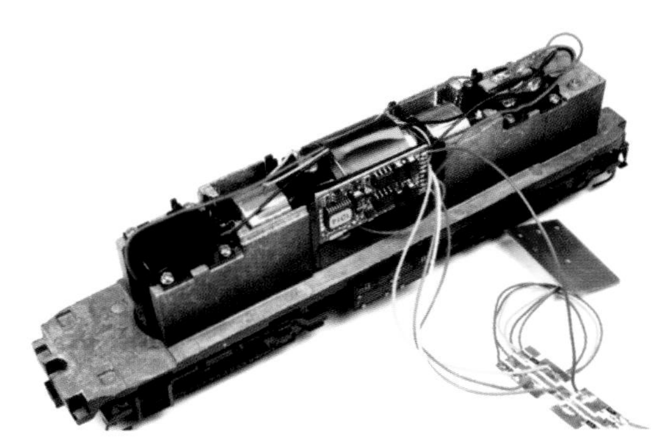

*The model is tested on a service track to ensure that the decoder has been installed correctly and the correct address readout obtained on the hand throttle. The default address for all brand-new decoders is 0003. The next stage of the project is to install the lighting kits. The main circuit board is visible in the foreground: note the coloured wires soldered to the circuit board so that each terminal can be correctly identified. Replace them one at a time with the corresponding coloured wire from the decoder.*

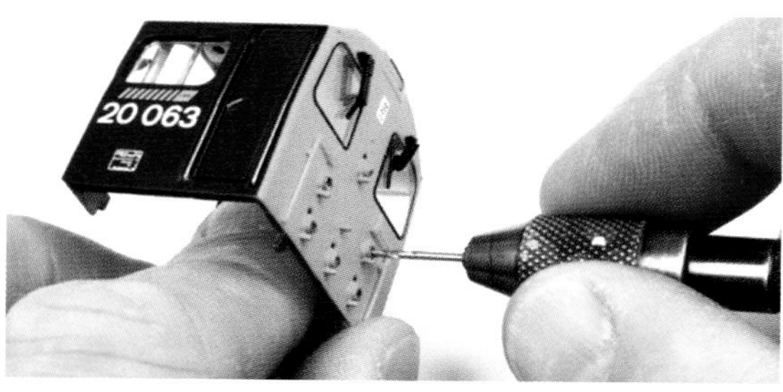


*The first model to be treated is the disc headcode type. This requires the drilling out of the required marker light lenses. It is good practice to drill a small pilot hole of about 0.5mm before opening it out to accept the optic fibre light guides provided in the lighting kit.*

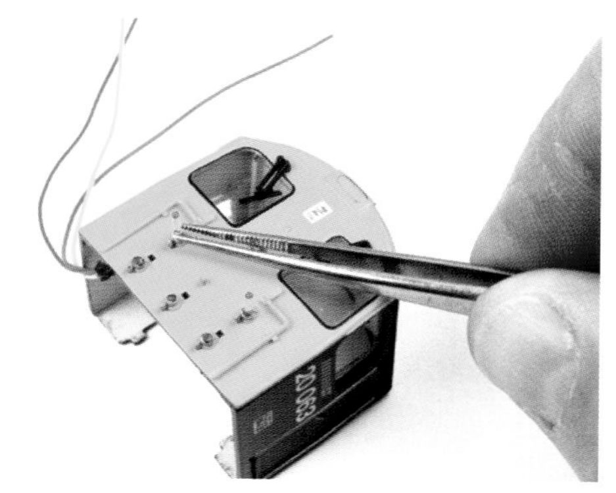

*Short pieces of light guides (plastic optic fibre) are cut to lengths of about 10mm and test-fitted to each marker light. Test-fitting is undertaken when the LED panel is fitted behind the cab front so that the correct length can be determined. Once in place any excess can be snipped off with sharp cutters. Follow up by cleaning the exposed end of each light guide with fine-grade wet and dry paper to give the impression of the marker light lens surface. The circuit is tested again to see that each light guide is transmitting a reasonable intensity of light from the light boards.*

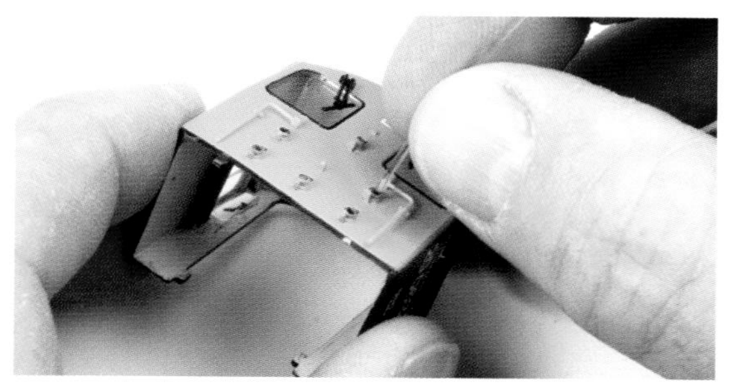

*Test fitting of the fibre-optic light guide.*

*This view shows how fitting the flat light board to the inside of the cab has a minimal impact on the detail provided by Bachmann. The light board can be fitted temporarily using Blu-Tac during testing and fixed with a more permanent adhesive later when the model is ready for re-assembly. The forward-facing control desk must be moved inboard by a few millimetres to clear the light board. Fortunately little other modification is required apart from drilling a hole through the cab bulkhead.*

*A discreet hole through the bulkhead provides a route for the wires to lead back to the decoder. The location of the hole ensures that wires are not seen through the cab windows of the model after re-assembly.*

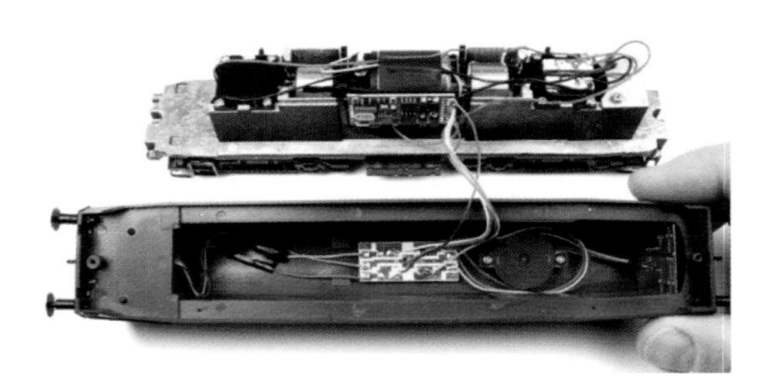

*Upon completing the project you should have an arrangement that looks something like this. Each component is carefully positioned to ensure that it does not foul on the bodyshell interior. One area to note is the cooling fan assembly, which can catch any additional circuit boards, wires or decoders. Electrical insulation tape is used to secure wires into place to prevent them from becoming pinched between the body and chassis frame.*

*Class 20s equipped with headcode boxes require a different lighting kit, one that includes LEDs to represent domino marker dots. The techniques are outwardly similar to the disc headcode version described above. Location of circuit boards and the decoder within the interior of the model remains the same.*

*The tip of a scalpel blade is inserted under the headcode box cover to prise it out gently. Turn it over and carefully remove the printed detail to leave it completely transparent. Put it to one side and replace once the fitting of the LEDs is complete.*

*The position of marker dots is carefully noted and a pilot hole of 0.5mm is drilled in each marker dot position. Once satisfied with the position, drill out with a 2mm drill bit.*

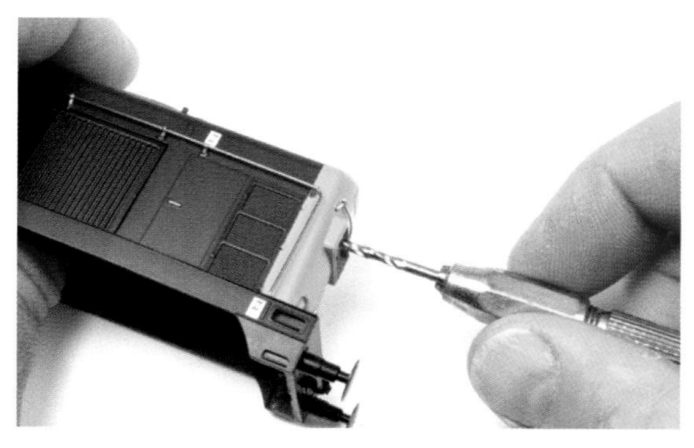

*Slowly does it! Time and care taken at this stage will ensure that your lighting components fit well. The headcode boxes should have black paint applied inside the frame, the drilled marker light holes and the interior face of the bodyshell. This will prevent light leakage and give the impression of a black panel with marker dots.*

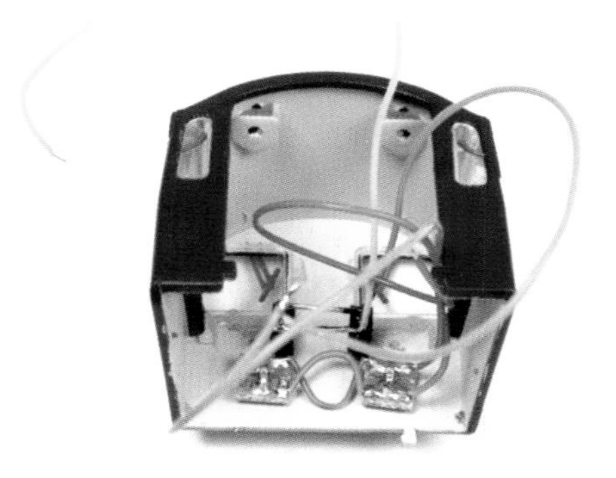

*At each stage of the project it is wise to test-fit the components to ensure that the fit is to your satisfaction. This image shows the components assembled to the interior of the cab. One control desk will have to be relocated back by a few millimetres to make room for lighting components.*

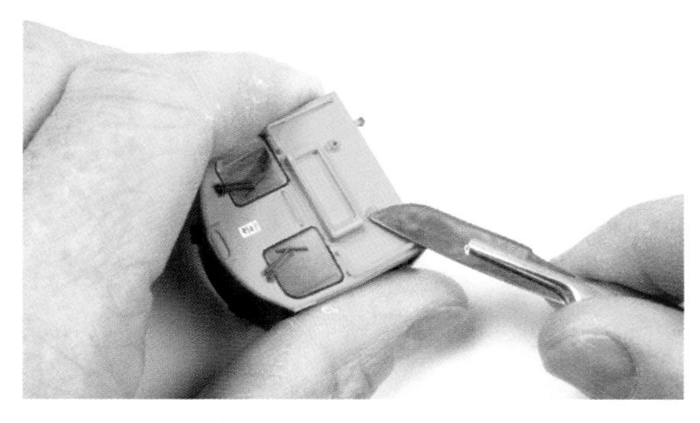

*Remove the tip from each of the tail marker lights before drilling holes to accept optic fibre light guides. This kit has the light guides pre-assembled to the circuit board that holds the red LEDs.*

The project is concluded by re-assembling the models and trying out on the service track once again before allocating a unique address to each model. The default function output to function buttons arrangement was retained, using functions A and B through button F0 to control the forward marker lights. The tail-lights were controlled independently by outputs C and D using buttons F1 and F2, respectively.

## INTERIOR LIGHTING AND BIPOLAR LEDS (HORNBY LIMA CLASS 121 DMU)

When a model is supplied with reasonable internal detail it suddenly becomes desirable to provide some interior lighting, and the Hornby version of the former Lima Class 121 single-car DMU offers a perfect opportunity to experiment with some internal electronics to take advantage of the function outputs on a four-function decoder. While the interior detail of the Hornby model is relatively basic, a blank canvas for detailing, in effect, it would not take much to detail it with seated passengers, discarded newspapers and some additional colours to add character to the interior. Once you have completed the detail modelling, it would be a shame not to illuminate it with a lighting circuit.

Another interesting feature of heritage diesel multiple units is the use of marker lights equipped with both forward-facing lights and tail-lights in the same housing. This offers an interesting challenge for the modeller to represent as accurately as possible using LED technology. There are two options. Firstly, it is possible to use a bicolour LED that, depending on which way the current passes through it, will emit either a yellow/white or red light. With careful wiring to decoders, bicolour LEDs can be easily switched to emit one light colour or the other, depending on the direction of the train.

The second option is to use tiny surface-mount SMT LEDs (*see* above). These are challenging to install but very effective when in use. A white and red SMT LED can be soldered to a small piece of copper-clad laminate circuit board to create a mini light board, which would be just small enough to fit behind the marker lights on the model, and wired in such a way that either a forward or reverse aspect is shown, depending on the train's direction.

Some modellers may prefer the SMT LED method, given that most bipolar LEDs tend to be rather large and emit yellow light rather than white as one of the two colours, which does not offer the authentic representation of incandescent lights that one might wish for. In addition, bicolour LEDs have three legs, of which the middle leg is usually a negative common, as opposed to two legs for a more conventional LED. For this project I decided to demonstrate the use of bipolar LEDs, because they have some interesting features and there is also the potential pitfall of the common negative leg to consider.

When choosing bicolour LEDs it is important to buy those with a positive common leg (or lead) and two negative legs, one of each leading to either side of the LED, relating to the individual colours. This matches perfectly the electrical arrangement for the function output on a DCC decoder that has a positive common wire, which should be connected to the positive common lead, and individual negative return wires for each decoder function, which can be connected to the individual negative legs of the bicolour LED. If you inadvertently select a bicolour LED with the opposite arrangement, it will not work with a decoder, so choose carefully and avoid this pitfall.

Once again, and in common with all the projects in this chapter, the techniques can be used and adapted to build internal lighting and light colour marker lights into any model of a heritage DMU. The techniques described are interchangeable with almost any project, provided that the rules for wiring of lighting circuits to decoders, including a resistor in series with each LED to reduce the voltage, are followed. Here's how I tackled the Hornby Class 121 DMU, a model equipped with a new power bogie and DCC interface socket, but still without any lighting circuits, which makes it an ideal blank canvas for a DCC lighting project.

*Essential supplies that should be to hand for the Class 121 project include fine 0.6mm equipment wire, which is easy to hide in the vehicle interior, bicolour LEDs, resistors, sticky pads and copper-clad laminate PCB.*

*In addition, a four-function decoder and light strip designed for coaching stock interior lighting are required.*

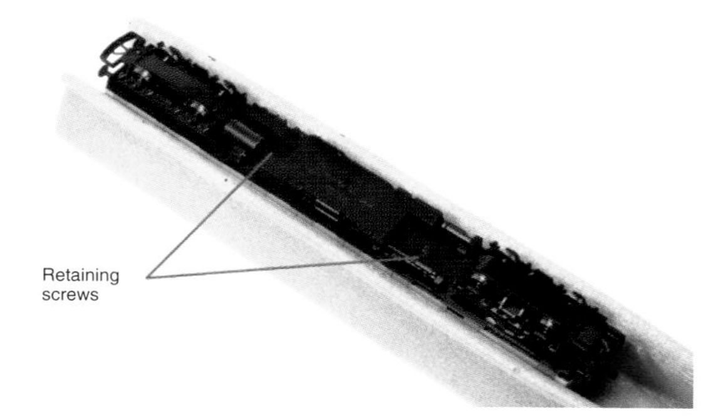


*Two retaining screws hold the chassis and body together. There are a couple of clips on the interior that should be prised away, too.*

*The interior lighting strip is tested before installation. Nothing out of the ordinary here!*

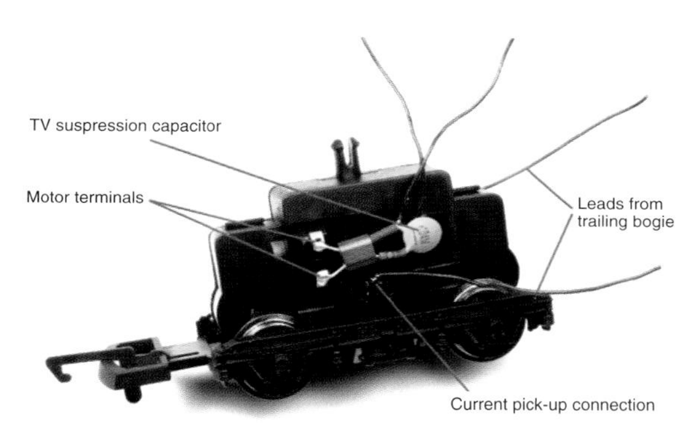


*Hornby has equipped its re-released Lima range with new drive bogies that are a considerable improvement over the obsolete 'pancake' ringfield drive motor.*

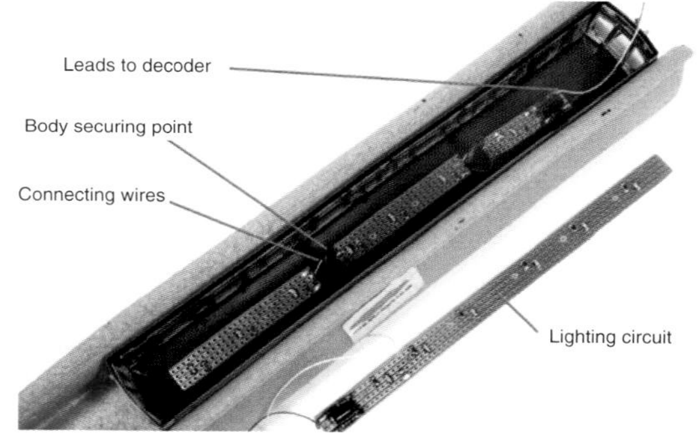


*Unfortunately the method of securing the body to the chassis means the model has two large screw fixings located right through the passenger compartment. They interfere with the lighting strip, which had to be cut into three lengths and connected together with equipment wire to complete the circuit.*

*LEFT: Dismantle the chassis until you have separated the drive bogie, interior seating and circuit boards from each other. This is the point when you first get an idea as to how the new lighting circuits will fit in the model. Multiple unit vehicles present the challenge of having numerous windows, all of which can cruelly expose your interior wiring.*

*The cab glazing inserts were carefully modified by cutting away the part that forms the light fitting lenses. If retained, they will only spray stray light throughout the cab glazing insert. The glazing plastic is brittle and cracks easily, so take care.*

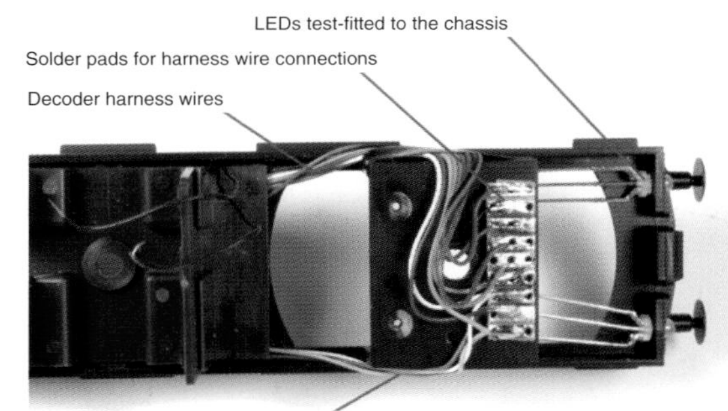


*Interestingly, there's room to hide the decoder under the interior detail moulding – quite a stroke of luck, considering how difficult it can be to hide a decoder in such an open vehicle. In the meantime, the chassis has been modified to accept the bicolour LEDs.*

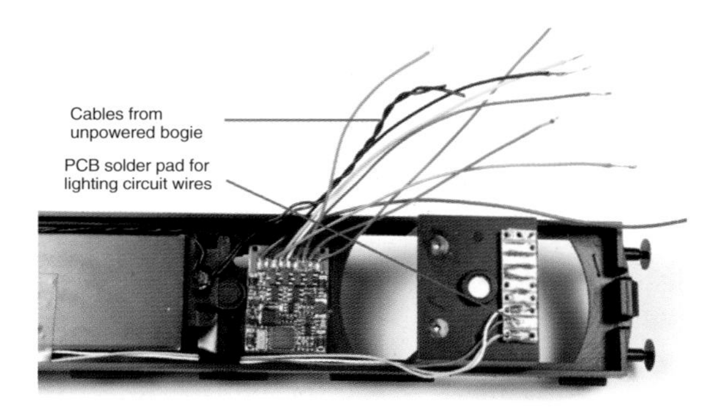


*The decoder is installed first. Note how each harness wire is connected to a mini circuit board located on the motor bogie bracket, which spans the bogie well. The circuit board provides solder pads to connect the various wires to the LEDs at each end of the model.*

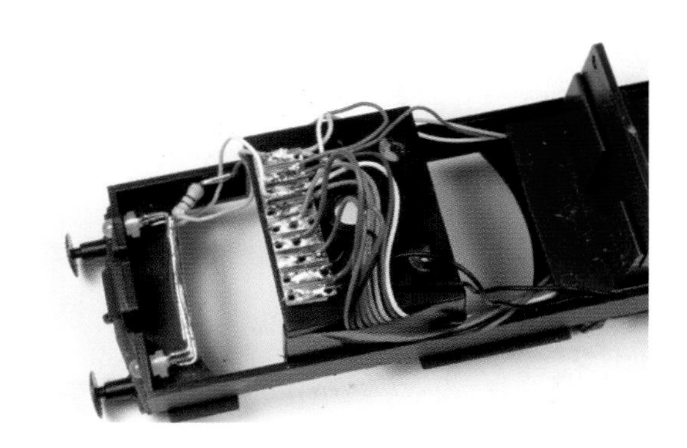

*The bicolour LEDs are connected to each other and to the circuit board placed on top of the motor bogie bracket. Note the use of a resistor on the blue positive common lead.*

LEFT: *Each of the four bicolour LEDs is to be fitted to the chassis frame, not the body. To locate the holes in the chassis frame correctly, refit the body and drill pilot holes with a 0.8mm high-speed twist drill through the holes in the marker light housings and through the chassis frame behind. Remove the body and open out those holes with a 3mm drill to accept the LEDs.*

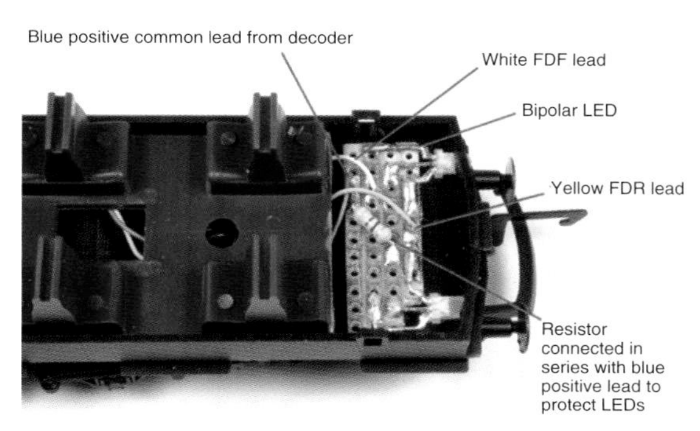


*At the opposite end of the model, the interior moulding has been trimmed to make room for a separate circuit board, which becomes the terminal board for the wires running the length of the chassis and a convenient connection point for the LED leads.*

*The driving cab desk has been reinstated, effectively concealing the LEDs.*

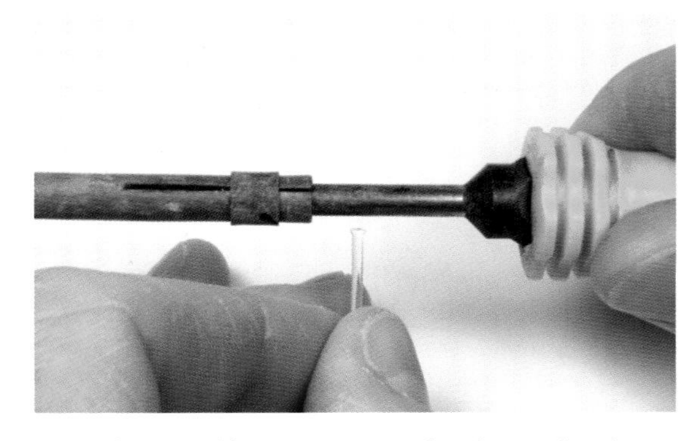

*Heat from a soldering iron is used to form a flat dome on a piece of plastic light guide. This becomes a new light lens once it has been trimmed to length.*

*Interior lighting is very effective in multiple unit stock. The tail-lights are also independently illuminated.*

*Interior detail is revealed, bringing the model to life with such features as passengers, litter and discarded newspapers.*

*The forward-facing marker lights are illuminated in this view. The head and tail marker lights share the same housing, an ideal application for bicolour LEDs.*

*Close-up of the forward marker lights.*

*Close-up of the tail lights.*

## LIGHTING IN MULTIPLE UNITS (BACHMANN CLASS 158/159)

The Class 158 model offered Bachmann the basis for producing other models of third-generation multiple units, including the Thames Turbo Class 166 (the proposed Chiltern Line Class 165 was dropped) and eventually the Class 170 Turbostar. The powered car in each model has a heavy die-cast chassis that was adapted for use for each class in the range. The underframe components are separate plastic mouldings that clip into the die-cast frame, making it easy to introduce detail changes. The drive mechanism and the bogie frames/towers could be used repeatedly on all of the classes offered.

Unfortunately for the DCC user, Bachmann has not yet upgraded this model with new circuits and NEM sockets that would enable straightforward conversion to DCC. Clearly there are practical difficulties with making older two-car and three-car multiple unit models DCC-ready as a retrospective upgrade, especially when the powered car in a three-car set is located in the middle. The lighting circuits must be adapted for DCC, too, with the modified circuits located in the unpowered outer cars. Clearly, for the modeller, the Class 158, 159 and 166 (together with the Class 170 Turbostar, which is outside the scope of this article because it is equipped with different lighting circuits) are going to be more difficult to convert to DCC than a locomotive. So, what are the options open to us?

*The beady-eyed look from the LEDs fitted to the Class 158 is particularly prominent on this Central Trains version of the model. They can be improved by filing the rounded surface of the LED lens flat with a fine-cut file – but carefully, so you don't breathe in the filing dust or damage the LED.*

*One method of supplying power from a decoder to the lighting circuits located in an unpowered vehicle is to fit cables, plug and socket arrangements between the cars. Soldering up the circuits together with the complexity of installing them within the gangways means that this is a fiddly and sometimes impractical solution that in the end does not mitigate the cost of the use of individual decoders in each multiple unit trailer vehicle.*

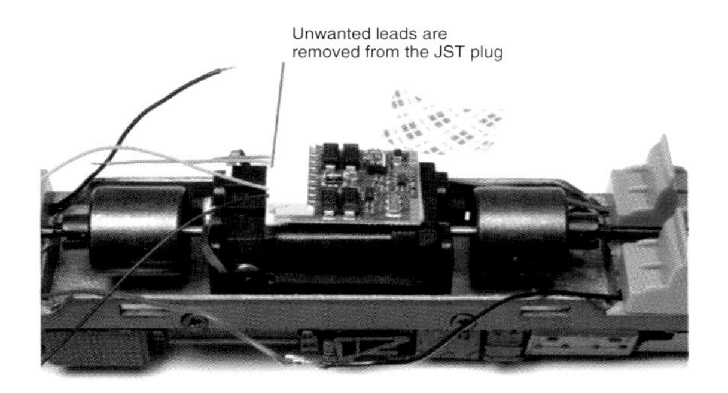


*The motor casing is covered in electrical insulation tape before the self-adhesive pad supplied with the decoder is stuck on top. This view shows the decoder in its final position and with unnecessary cables removed from the wiring harness plug to leave only the red, black, grey and orange cables.*

*This is the arrangement once the lighting circuit has been removed from the cab. This particular model was fitted with current collection pickups at one end only. Current collection could be made more reliable by making up a second set of pickups from phosphor bronze strip and PCB for the inner end bogie.*

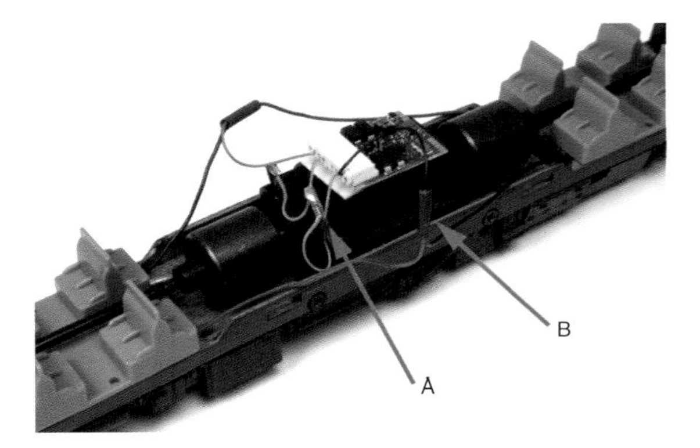


*Installation is complete with the black and red wires from the harness connected to the track supply cables. The orange and grey cables from the harness are directly soldered to the motor terminals. Exposed soldered connections are protected with heat-shrink sleeve to prevent accidental contact with the decoder (not a good thing) or the chassis frame. This is a good example of a simple DCC hardwire installation where no NEM socket exists.*

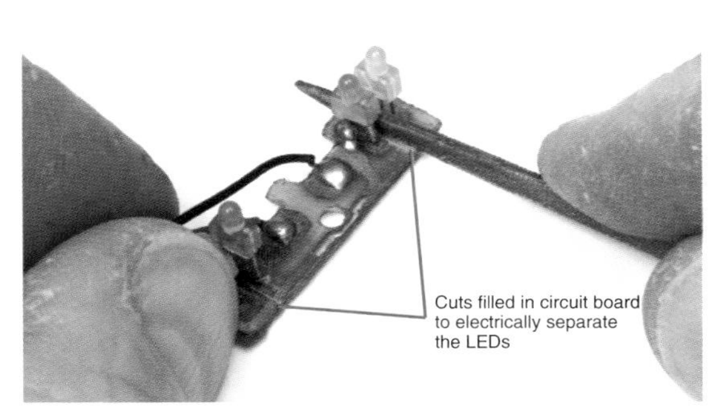


*Two cuts are made in the circuit board to isolate the paired red and yellow LEDs from each other. Note the positive pole of each, marked on the circuit boards.*

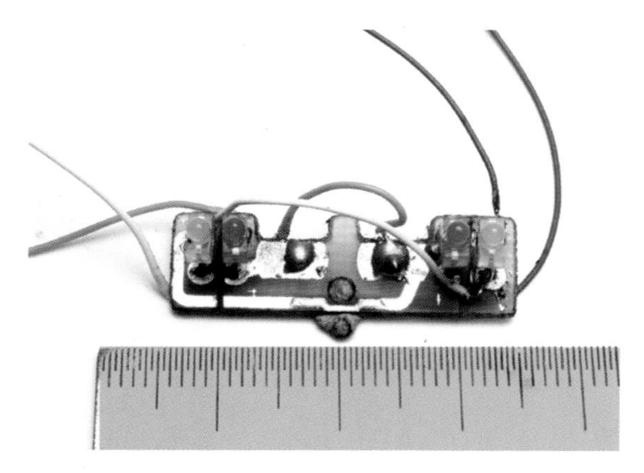

*The completed changes to the circuit board can be seen in this view. The blue cable represents the connection that will be made to the positive common output of the decoder. This cable is connected to the positive pole of the LEDs. As this lighting circuit will be returned to the outer car that will represent 'forward', as far as the decoder operation is concerned, the yellow LEDs representing headlights are connected to the white cable on the decoder. The yellow cable is connected to the red LEDs' negative pole and is connected to the yellow cable on the decoder.*

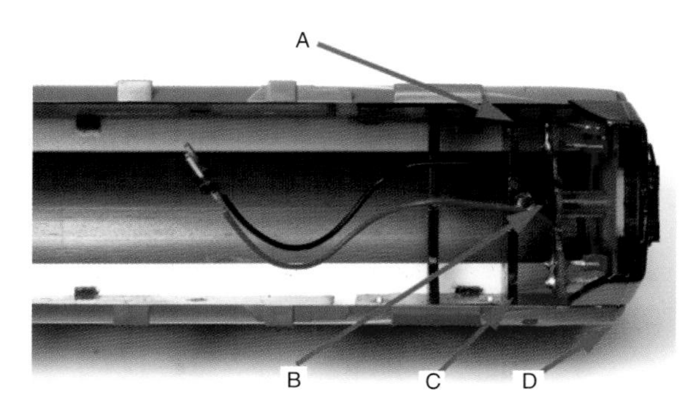


*Once inside, it is easy to locate the simple circuits at the cab end of the bodyshell. A plug and socket arrangement means that the body can be separated from the chassis. To modify the lighting circuit, it is necessary to release the retaining screw at B. Unfortunately, the method of assembly used by Bachmann means that a section of glazing that incorporates a cross member within the bodyshell is in the way. You have the option of cutting this away by making cuts at A and C using a piece of hacksaw blade. Alternatively, the glued join between the cab and the body could be 'released' (broken) so you can reach the lighting circuit retaining screw.*

CHAPTER 6

# Digital Sound and Steam

*Digital sound is becoming mainstream thanks to developments by such manufacturers as Bachmann, Athearn and Broadway Limited Imports. One of the first UK-outline ready-to-run locomotives to be adapted to host a digital sound decoder and speaker was the Bachmann Class 20. Modellers were previously limited to buying decoders and speakers independently. This left the modeller to determine how to build it into a favourite model, sometimes using guesswork, trial and error for a satisfactory result.*

As the boundaries of what can be done with DCC are pushed back by the modeller, special effects soon come to the fore and none are more evocative than the sight of a locomotive in steam or hearing the sound of your favourite diesel locomotives hard at work. In fact, O gauge modellers can enjoy fan-assisted exhaust in their diesels and digital sound is also available for steam locomotives in many scales, too, with the speaker placed in the tender or coal bunker when space is tight within the locomotive.

The popularity of digital sound has prompted

the mainstream manufacturers in Europe and the USA to develop models with authentic sounds, off-the-shelf and at very reasonable prices. Some of the products are rudimentary and some are highly sophisticated. None the less, they offer a new dimension to railway modelling, which comes to the fore when operating a layout. Digital sound brings in another of the senses, hearing, to complement the visual appearance of your models. For some model railway operating enthusiasts, if it sounds right, it is right.

Before we get to the noisy bit, let's take a look at generating some rising damp!

*The interior of the Bachmann digital sound Class 66.*

*Externally, you would have no idea that there's a sound decoder and speaker in this model. That typifies DCC – most of the technology is hidden from view until you press button F1.*

*Bachmann was not the first to fit digital sound decoders to a Class 66. An HO version of the locomotive as used in mainland Europe was offered by Mehano with the option of digital sound.*

*Another Bachmann model with digital sound off the shelf: the EWS Class 66. When it was originally released, it retailed for between £118 and the list price of £157 – not at all bad for an out-of-the-box digital locomotive with very convincing sound.*

*Factory-fitted sound decoders and speakers make life so much easier for the modeller. This is the interior of the Mehano HO gauge Class 66.*

*For most locomotives, a do-it-yourself approach has to be taken. Thankfully, there are various sound decoders available supported by different types of speakers and enclosures. Speakers are available separately, too. This photograph shows an ESU decoder fitted with the standard 23mm-diameter speaker. Adjacent to it is the larger 26mm-diameter round speaker together with an oblong 20 × 40mm version.*

*Fancy recreating something resembling the exhaust from A4 Pacific No. 60009 'Union of South Africa' as it powered south down the Highland Mainline in April 2007? Steam effects can be created with steam generators by Seuthe and other manufacturers, powered by the function outputs of a decoder for constant steaming and independent control for when the locomotive is parked out of steam.*

*Seuthe steam generators come in a variety of sizes to suit differing applications. A popular choice is No.22, the universal type for plastic locomotives with narrow chimneys. Some are specifically designed to work with decoders.*

## HOW TO FIT A SMOKE-EFFECT GENERATOR

Smoke generators have been available for years in various guises, the most successful being produced by Seuthe, which offers different types to suit the various situations that modellers will encounter, given the wide variety of steam locomotive types. One of the more useful steam generators is the Seuthe No.22 universal generator, which has a narrow body to fit slim chimneys and is suitable for use with plastic bodyshells. This project uses the Hornby A4 Pacific locomotive as an example of how to tackle this type of project.

*Two different versions of the Hornby A4 Pacific locomotive have been chosen for this project. No. 60031 'Golden Plover' is equipped with a 'double' chimney. The model is superbly detailed with many delicate standalone parts. The chassis is equipped with a NEM socket, which means it can be easily converted to DCC using a decoder equipped with a NEM eight-pin plug.*

Modellers must ensure that the power consumption of the chosen smoke generator does not exceed the maximum rating of the decoder function output, which may be as little as 100mA. Suggestions of combining two function outputs to double the current available to the smoke generator can also lead to an over-voltage situation: two outputs will double the voltage to 24V dc, sufficient to cause overheating of the generator, which may be rated for a maximum of 16V. This could be overcome by turning the output voltage down for both outputs if there is a CV setting that permits such a thing. Most decoders offer a 'dimming' feature for outputs when used with lamps or LEDs and that is the decoder function CV you should use. Should you decide on this option, wire both the negative wires from the chosen function outputs to one lead of the steam generator. Programme the decoder so that both functions are activated at the press of a single function button.

Another consideration will be the amount of room to accommodate the steam generator. The A4 Pacific used in this project had to be modified to accept the Seuthe No.22 generator by filing away a corner of the chassis block. Some locomotives may need drastic surgery to accommodate a steam generator; be prepared to file or cut away a great deal of metal in some instances.

*'Guillemot' is the second of the A4s chosen for this project. It is interesting to place both models side by side and compare the detail differences applied by Hornby. 'Guillemot' has a single chimney, which requires a slightly different approach to fitting the smoke generator.*

The first part of the conversion process was to remove the chassis from the body. This must be done carefully with some models because parts of the running gear, such as the speedometer drive cables, may be located between the valve gear and plastic bodyshell. The decoder was fitted first by simply removing the dummy plug from the NEM 652 DCC interface socket in the chassis and plugging in the decoder. The position of the No.1 socket is clearly marked on the circuit board and the decoder plug should be aligned so that the orange wire and pin is plugged into that socket. I placed black insulation tape on the area of metal chassis before using the supplied sticky pad to secure the decoder. The locomotive was tested on the service track to ensure there were no faults before installing the steam generator.

The blue, white and green harness wires were disconnected from the decoder plug so that they could be soldered to a separate circuit board made up from a piece of copper laminate PCB. This provides solder pads for the connection for features such as LED 'oil lamps' and steam generator. The yellow and purple wires were removed from the harness as these were not required.

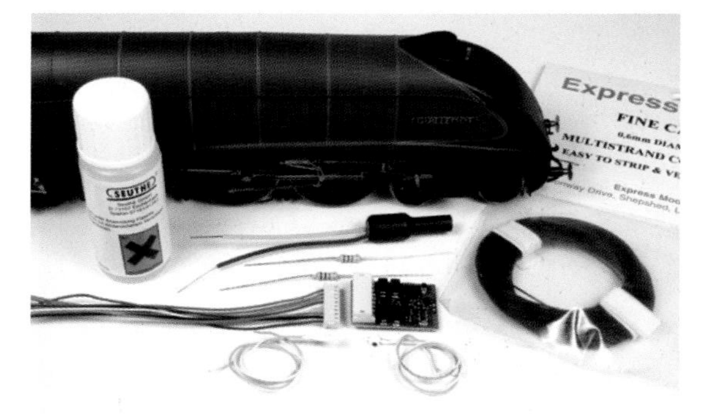


*A No.22 Seuthe steam generator was chosen because it is regarded as the best match for this type of model. Note that the Lenz Gold decoder is equipped with a cream-coloured socket into which the harness is plugged. This is referred to as a JST socket* (see *Chapter 4). In this project, having a JST plug to enable removal of the decoder without disconnecting the harness wiring is a distinct advantage.*

Normally, the removal of harness wires would invalidate a decoder's warranty. However, the Lenz Gold was chosen for this project because it is equipped with a JST socket that accepts a harness fitted with a JST plug (as opposed to a harness that is soldered directly to the decoder). This means that the modeller can remove the decoder from the locomotive at any time without having to disconnect the harness from the model and that a decoder can be refitted with a new harness without any soldering should it be required for use in a different model. It also means that you can remove unwanted harness wires that would take up valuable room in the model without modifying the decoder and invalidating its warranty.

The steam generator was inserted into the chimney and the plastic body refitted to the chassis so that the optimum position of the generator within the chimney could be determined. After removing the body once again, the generator was glued into place with an epoxy-type glue such as Araldite. The wires protruding from the bottom of the steam generator were soldered to the circuit board so that one wire was connected directly to the green decoder harness wire and the other was connected to the blue harness wire (the positive common supply).

When attempting to reassemble the first of the two locomotives, it was discovered that the rear chassis clips would not fully engage with the plastic bodyshell. The steam generator just caught part of the chassis frame at the front of the locomotive. This was resolved by filing away a small corner of the metal chassis frame to make room for the generator and to allow the body clips to be manoeuvred into place.

*One large screw located at the front of the model and located underneath the front pony truck secures the body and chassis together – it is easily accessible. The rear of the chassis clips into the body adjacent to the cab, which means that the chassis has to be rotated out of the bodyshell. Ensure that the speedometer drive cable has been disconnected first.*

*The speedometer drive cable is secured to the valve gear with the same hexagonal nut that secures the coupling rods. All of this detail is very delicate and should be treated with care, in case the valve gear is bent or any parts come loose.*

RIGHT: *Unlike some older models of steam locomotives, the new Hornby A4 is fitted with a NEM 652 DCC interface socket, making it simple to fit a decoder. Furthermore, there is plenty of room for a 1 amp HO/OO decoder such as the Lenz Gold used in this project. The dummy plug is removed and the decoder plugged in, with the plug aligned so that the orange wire pin is plugged into socket No. 1. This is indicated on the circuit board with an arrow.*

Each locomotive was tested on the service track for a second time to see that no faults were detected and the steam generator did not constitute an overload that would be detected as an error. Then straight on to testing the models on full track power. When this project was undertaken I was keen to see how these steam generators would work. In the event, they were quite spectacular, filling my working area with a dense white smoke in no time at all.

*The Seuthe No.22 universal steam generator is a compact unit that fits the chimney of an A4 nearly perfectly. It is glued into place with a five-minute epoxy glue such as Araldite 'Rapid', which is strong and stable. There is sufficient room at the front of the chassis between the chimney and the chassis block to accommodate this type of steam generator and its electrical leads.*

## OPERATING LOCOMOTIVES EQUIPPED WITH STEAM GENERATORS

Following further testing of the project locomotives on the service track, the steam generators were put to the test. Each generator is packaged with a small quantity of 'smoke oil' and dispensing syringe. Further supplies of smoke oil can be obtained from Seuthe stockists, in quite

*This is the view of the steam generator after fitting to the chimney of 'Guillemot', the single-chimney version of the A4. This is as far as it is possible to recess it inside the chimney without resorting to removing parts of the chassis block immediately underneath the steam generator.*

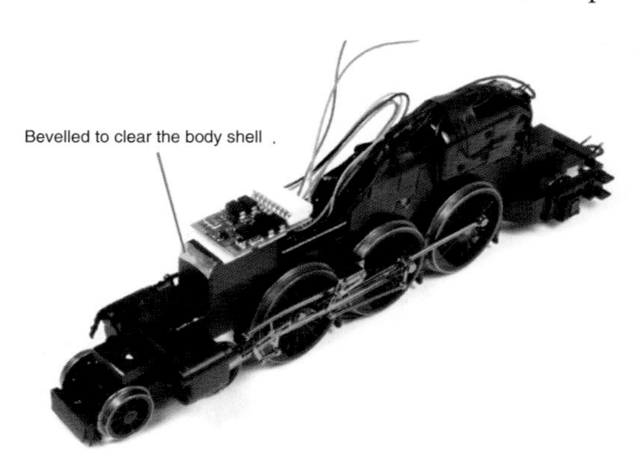


*Before any attempt was made to fit working representations of working oil lamps and the Seuthe steam generator, the model was first tested and then equipped with the decoder. Three of the decoder wires are disconnected from the NEM plug – blue, green and white. These will be used to power the LED 'oil lamps' and steam generator and will be soldered to a separate circuit board, which is isolated from the model's main wiring.*

*Note that a small piece of the chassis block has been filed away as indicated by the red arrow. When the decoder and steam generator are fitted to the model, it is impossible to refit the body to the chassis unless this small corner of metal is removed.*

*'Golden Plover' in steam on the photo stage. The effect is most impressive and a couple of drops of oil last for a reasonable period of 'steaming'.*

*Digital Command Control offers the same independent control of a steam generator as it does of lights, even when the locomotive is stationary. It would be impossible to obtain this photograph if the steam generator were operated with 'analogue' control without disconnecting the motor leads.*

*To operate a Seuthe steam generator, oil is injected carefully into the chimney and the generator switched on by pressing the appropriate function button. There is a tiny element in the middle of the generator that should be treated carefully as it is most delicate. Do not touch it with the tip of the needle.*

large quantities if the smoky atmosphere of a steam shed is your objective. Following the manufacturer's instructions, a small quantity of oil was drawn into the syringe and injected into the top of the steam generator, a few drops at a time, taking care not to touch the central heating element.

The steam generator was wired to the green harness wire of the decoder, which means that it would operate from function button F1 if default function output control is retained. A few seconds after pressing the button, white smoke was emitted from the chimney with notable force, as if it were driven by pressure. The effect is surprisingly effective. When run on the service track, the 'steam' billowed behind the locomotive in a fairly authentic manner. On a personal note, I must confess that I had been sceptical about 'scaling down' steam to a scale as small as HO/OO gauge, but the effect was none the less impressive.

The only practical downside of the use of such a generator and smoke/steam oil is that it does tend to splutter a little when it's first switched on, spraying a small quantity of oil onto the bodyshell of the model. Fortunately the oil has been formulated not to affect model paintwork or plastic. When the model is removed from the layout for storage or transport, any remaining oil in the generator should be removed. It also occurred to me during testing of the locomotives that the full 12V output might not produce desirable results for some locomotives and adjusting the output voltage of decoder output C in the CV for setting 'brightness' will reduce the amount of steam that is produced and the force of the exhaust. Some experimentation is always useful to achieve the optimum effect, bearing in mind that most generators of this type are designed to work at 12 to 16V.

## DIGITAL SOUND

As many of the favourite locomotive classes have disappeared from the railways in favour of uniform-looking Class 66s, or large wide-cab diesels in the USA, digital sound is growing in popularity for re-creating the sounds of once numerous classes of locomotives that were so often taken for granted. It's a third dimension that takes you straight back to the linesiding of years gone by, when less onerous sound and emission laws meant that locomotives were noisy and generated some real exhaust 'clag' to go with it, too!

There are several options available to modellers when it comes to digital sound. In recent years the mainstream manufacturers have developed ready-to-run locomotives with sound decoders and speakers factory-installed so that the modeller can enjoy digital sound straight out of the box. However, not all classes of locomotive are available with digital sound and sometimes it is left to modellers to locate and install their own sound decoders. In some instances, modellers will look for a 'best fit' option and select pre-programmed sound decoders currently available from DCC manufacturers such as Digitrax, ESU, SoundTraxx and ZTC. Fortunately there are a number of after-market specialists who go to a lot of trouble to obtain authentic sound recordings from real locomotives, spend an extraordinary amount of time splitting the sound recordings into digitized sound packets and then install them into the sound memory slots in sound decoders, making them available as bespoke sound decoders for specific classes of locomotive.

Steve Weeks of South West Digital is a pioneer in this area in the UK and should have the appropriate recognition, even though other after-sales companies are offering their own versions of sound decoders. His company continues to offer sound decoders with bespoke sound carefully digitized and programmed into ESU decoders for locomotives such as the Bachmann Class 08 shunter, the Class 47, Class 40 and Class 37, together with Hornby models such as the Class 31. The company continues to develop sound decoders for many other classes of locomotive and these are generally available at about the same price as an 'empty' digital sound decoder from ESU. In the USA, sound decoders with bespoke sound for EMD and GE locomotives are available from a number of after-sales developers, usually specialist DCC retailers such as Litchfield Station of Avondale, Arizona.

So, your favourite model can be equipped with a proper representation of the real locomotive's sounds, everything from warning horns, whistles, bells and, of course, the engine or cylinder beat itself. But how does a sound decoder do that?

Digital sound decoders have additional memory for storing packets of sound, which are prepared by careful work with an original recording. Those packets have to be restricted in size because the memory in a decoder is limited. Looping of some sounds has to be done to achieve engine beat for a diesel locomotive and the skill of the programmer should prevent the result from sounding 'spooled' when the model is in operation. Having tried it for myself, it's not easy.

The memory is divided into a series of timed slots, which are occupied with certain types of sound packet. Random sounds are those representing equipment on a locomotive that trips in at certain intervals, not under the control of an engineer or driver. They include compressors, cooling fans, motors and air dryers. These sound packets occupy certain timed slots in the decoder memory. Then there are operator-controlled sounds that would be the same as those activated from the cab by the driver: horns, bells and whistles, for example.

Finally, there are slots holding packets of sound that are controlled through the throttle as a locomotive is run, including changes to the engine beat as a locomotive is powered up or throttled back to idle for coasting. Such sounds are activated as the decoder progresses from slot to slot as the locomotive proceeds along the track. When running at a constant speed, the decoder plays the same sound from the appropriate slot, over and over in a loop. As the locomotive is driven those sounds change automatically according to the throttle setting as the decoder moves from slot to slot and plays the sound placed in the slot. In one sense they are operator controlled by pressing the starter button to start the engine at the beginning of an operating session and shutting it down after the session with that locomotive is over.

The sound is amplified by the sound decoder so it can be played at an audible volume through the other piece of hardware, the speaker. Finding a suitable spot to locate a speaker together with its enclosure within the confines of the model is usually more challenging than locating the decoder itself. While the practice for sound decoder installation is no different to that for the fitting of 'silent' mobile decoders, speakers are an entirely different matter.

As is frequently the case in our hobby, space is a limiting factor. Whereas small speakers will better fit the tight confines of our typical locomotive models (O gauge modellers are at an advantage when it comes to interior space), sound reproduction quality will improve the larger the speaker that can be installed. Use the largest speaker that will fit your particular locomotive or alternatively try installing two smaller speakers, which will enhance the volume, if not necessarily the overall sound quality. When space is tight it is very tempting to dispense with the baffle or speaker enclosure, which can take up most of the room, as you will see in the sample projects below. I must confess to making this mistake when installing my first sound decoders, although the sound quality was not adversely affected because the locomotive body itself did something to actually amplify the sound. Whenever possible, you should always use the speaker enclosure supplied with the speaker itself, because it has been specially designed to suit it. The materials selected for speaker enclosures will affect the quality of sound reproduction, therefore you have to assume that the manufacturer has chosen the optimum material to work with their speakers.

When space is tight, however, you have several options. You could, for example, consider manufacturing your own speaker enclosure from styrene, making it bespoke to suit your model's particular characteristics; I did this to an early Bachmann Class 37 with good results. You could also cut down the speaker enclosure supplied, provided there is sufficient room between the back of the speaker and the bottom of the modified enclosure to enable the speaker to perform at its best.

So, why are speaker enclosures so important anyway? The reason is simple. A speaker produces a forward and then a reverse sound wave, but if the baffle is not present or there is no enclosure made of styrene card or other material to provide an airtight seal between the front and the back ways, the sound waves will cancel each other out, producing a poor sound level. Speaker enclosures have been designed to optimize the volume output and, given the sizes that we are limited to using, speakers need every bit of help they can get.

An interesting point that modellers should consider is that it is not always necessary to let the sound 'escape' from the bodyshell. It is tempting to mount the speaker immediately below the fan grille or other similar opening. However, I have experimented with locating speakers within the enclosed bodyshell of several locomotives and have found that, by locating it in a position where it does not touch the bodyshell, causing vibration or resonance, and by enclosing it in a large space, the sound is actually amplified and the bass is improved (the small speakers generally used with sound decoders are not particularly good at reproducing the bass sounds so characteristic of many diesel locomotives). There is little doubt that the installation of digital sound in small models, particularly when modelling in N gauge, can be quite challenging both in respect to the location of speakers and achieving the right sound reproduction.

A further dimension to digital sound and your model railway can come from finding or making a sound recording of your favourite locomotive and working on it yourself in order to program digital sound decoders. Companies such as ESU offer sound decoder programmers that can be used in conjunction with a personal computer and the appropriate software. This works on the assumption that you can obtain your own sound recordings, digitize them and divide them into sound packets that are sufficiently small to fit in the sound decoder memory slots. While computer enthusiasts may find this an appealing project to attempt, it can detract from the main objective, which is to build a model railway. Ultimately it can take considerable time to program a decoder with digital sound correctly. Sometimes you may find yourself actually reinventing the wheel because a bespoke sound decoder for that particular class of locomotive may already be available at a reasonable price.

A couple of cautionary notes should be considered when looking towards digital sound for your model railway. Some modellers feel that buying an expensive sound decoder for each of their locomotives is excessive and sucks modelling budget away from other areas of the layout, such as building structure kits and creating scenery. This is absolutely correct, since to equip every locomotive on a large layout with a sound decoder would be pretty pointless: the overall volume produced from even a few of them would be irritating and would soon become featureless. Many modellers prefer to install digital sound in a select number of locomotives to create the right atmosphere, using just one sound-equipped locomotive in every consist of three of four used on a large freight train, for example.

While on the subject of sound reproduction and particularly its volume, most digital sound decoders, if not all, are supplied with the volume set to maximum as a factory default. When the decoder is set to maximum volume, it can distort the quality of the sound as well as push the performance of the amplifier within the sound decoder to its maximum. In extreme cases this can result in the decoder overheating, which will trigger the thermal cut-out device. The volume control in ESU decoders is found in CV63. The maximum volume setting, which is also the default setting of the decoder, is 64. I found that turning it down to a value of 50 provided the optimum sound quality for use at home. It is surprising how loud sound decoders can be when used in the domestic environment. At exhibitions or train shows, however, they are often drowned out by background noise and that is the time when it may be necessary to turn up the volume. It is a good idea to do this by using 'programming on the main' because, if you wish to demonstrate that to an observer, the change in volume level is instantaneous when the new CV value has been accepted by the decoder!

Another dodge considered by many modellers is to purchase a single sound decoder and install it, complete with speakers, under the layout rather than in the locomotive itself. One of the major benefits of installing the speaker within the confines of the locomotive model is that the sound source travels around with the locomotive itself, as it would in a full-sized machine. This is infinitely more authentic and draws the eye to the movement of the locomotive and its train. Also, the key to digital sound is being able to operate specific sounds, such as bells and horns, according to the situations found on the model railway as the operator drives the train, in exactly the same manner as a locomotive engineer or driver would when driving on the full-size railway. This brings us back to the point I have emphasized throughout: everything we do to make DCC work on our model railway is to enhance the driving and operating experience and make it as realistic as possible.

## DIGITAL SOUND PROJECTS

The following projects demonstrate different methods of installing a digital sound decoder and speakers in a variety of situations. Some models are large enough for two speakers for optimum sound reproduction. Some require ingenuity to squeeze the speakers into the model. Others again require extensive surgery to the inner structure just to create the required space. Each project can be used as reference to overcome the challenges of other types of model.

## DIGITAL SOUND IN A HELJAN CLASS 47: TWO LARGE SPEAKERS IN ONE LOCOMOTIVE

The Heljan Class 47 is generally regarded as a particularly good model despite reports of inaccuracies in the overall width of the bodyshell. Modellers appreciate its good running characteristics and the performance of all the examples in my personal collection has been consistently good. While Heljan models tend to be a little greedier in power consumption than those from other manufacturers, a quick check with an ammeter will soon tell you if the locomotive you have chosen to receive a sound decoder will exceed the decoder's maximum rating. Assuming all is well, when you take the body off you will see that the design of the chassis lends itself to the fitting of a sound decoder and speakers. The considerable room available permits the use of two speakers: the 23mm diameter speaker normally supplied with an ESU decoder and perhaps also the larger 26mm diameter model. This project shows how I installed two speakers and decoder in one of my Heljan Class 47 models. Very little modification was made to the locomotive itself. The speaker baffle supplied with the 26mm diameter speaker was trimmed slightly so that it would clear the shoulders of the locomotive body.

*A Heljan Class 47/3 dressed in Railfreight grey livery. It has been weathered and detailed as part of its makeover. Sound adds the operational detailing dimension to the visual appearance of the model.*

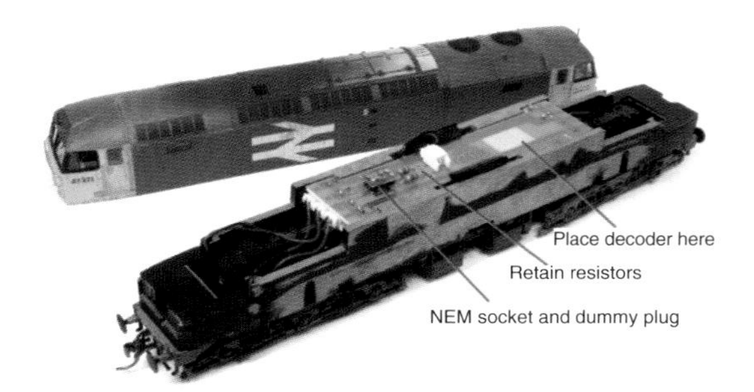


*The interior arrangement with dedicated space for the decoder located on top of the circuit board, DCC interface socket and lighting circuits.*

*The decoder is fitted first and then tested on the service track. A sticky pad is ideal for securing the decoder.*

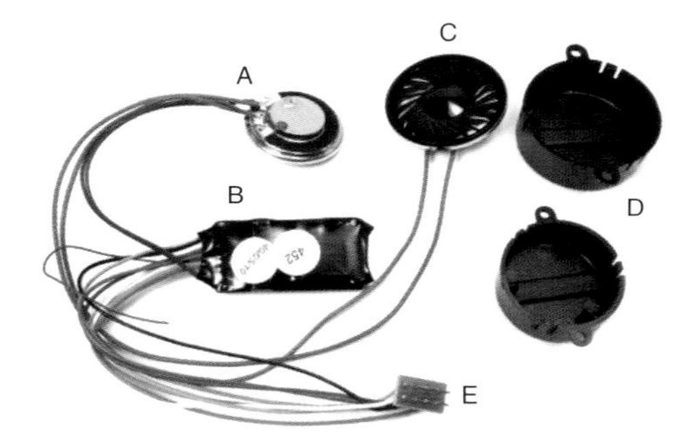


*The chosen decoder is produced by ESU, together with two speakers wired in parallel (the 26mm-diameter speaker is purchased separately):*
*(A) 23-mm diameter speaker supplied with the decoder; (B) decoder; (C) 26-mm diameter speaker; (D) enclosures; (E) NEM interface plug.*

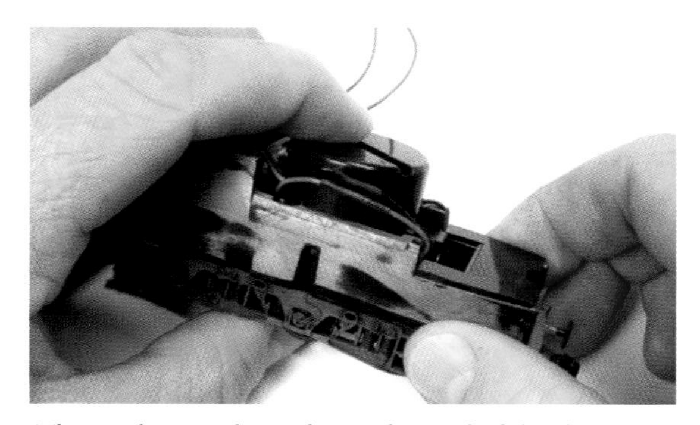

*The speakers are located on either end of the chassis frame, facing down towards the bogies. If they were placed facing towards the roof, securing them to the chassis would be that bit more difficult.*

*The larger speaker's enclosure is modified by bevelling the edges so they do not come into contact with the interior of the bodyshell.*

*Each speaker is secured with strong electrical tape applied over the back of the speaker enclosure and down the side of the chassis frame.*

*It may look slightly insecure, but using tape works well. It also means that the equipment can be easily removed for repair or modification, if required. Fortunately there is sufficient clearance between the top of the bogie gear tower and the speaker for the bogie to move around as it should.*

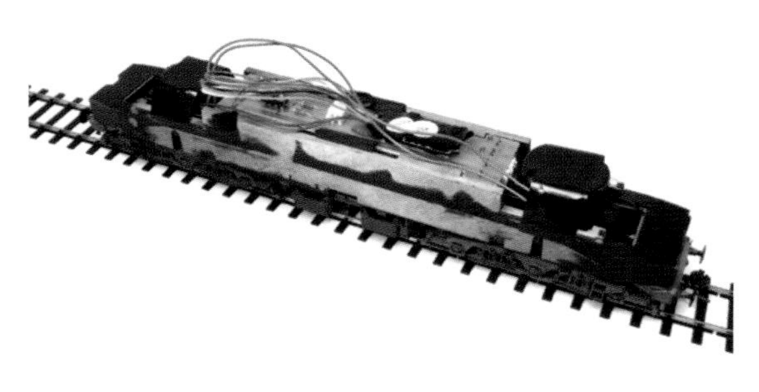

*The completed project showing the extension wires linking the two speakers to each other.*

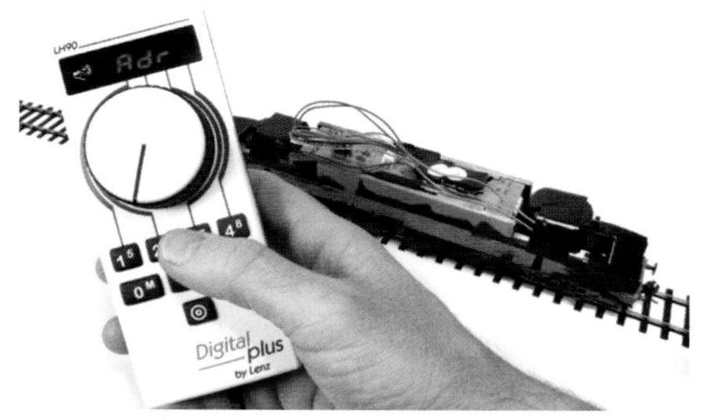


*The installation is tested on the service track to ensure the command station does not detect any faults. After a unique address has been chosen for the locomotive, it may then be placed in service.*

*That's it, job done. Externally nothing suggests that a piece of digital equipment, worth more than the model when bought new, has been concealed inside. Now press function button F1 and see what happens!*

## BACHMANN CLASS 57: USING THE FUEL TANKS

When space is tight – so tight that a speaker just will not go in – you must hope the fuel tanks have not been filled with die-cast metal, too. In the case of the Bachmann Class 57, the tanks are remarkably free of clutter, making installation simple.

*The chosen project model was the First Great Western Class 57/6 version of the Bachmann Class 57, a fine-looking model enhanced with sound effects as well as its operational running lights.*

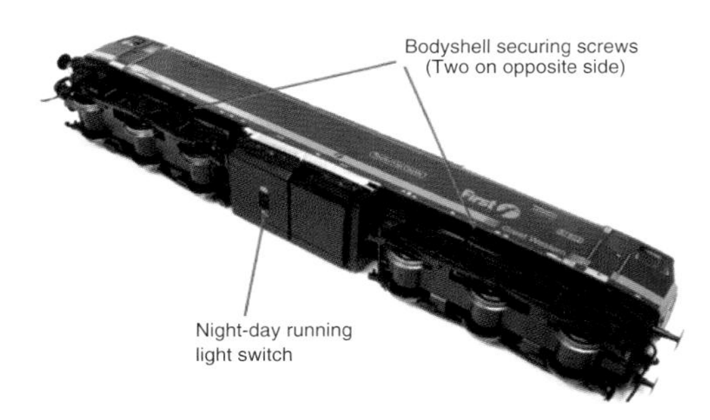


*Bachmann uses screw fixings to secure the body to the chassis, including one at each end under the cab.*

*The cab end screws are released first using a cross-head screw driver. The four central bodyshell screws are released next and caught as they drop out of the chassis.*

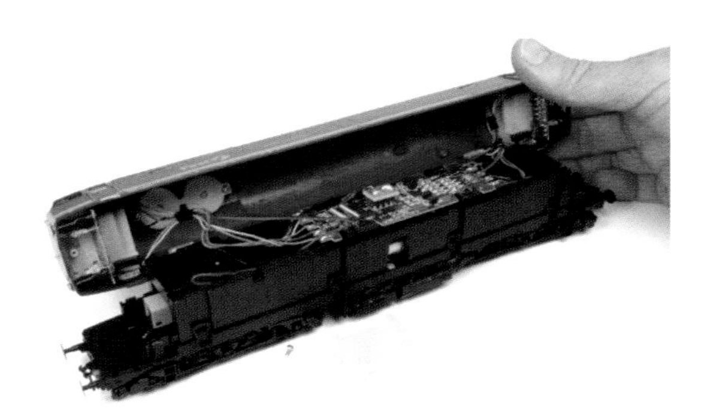

*As you can see, there's not a great deal of room for a speaker. Subsequent releases of this model have been improved by rearranging the lighting boards and reducing the amount of wiring between the chassis and body, although the actual chassis frame remains essentially the same. In this case, however, we have to contend with the wiring between the body and chassis.*

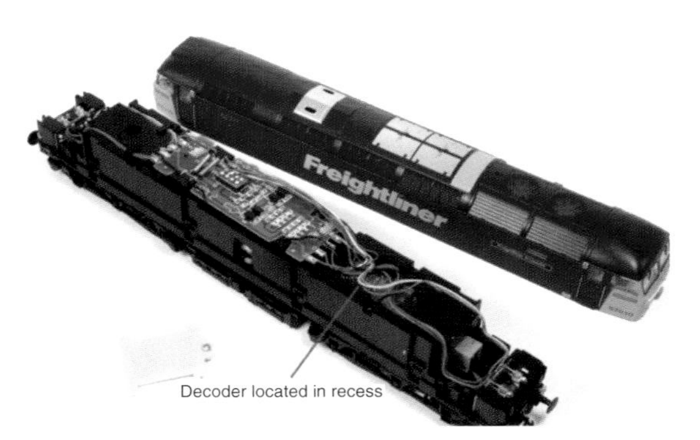


*This view of the Freightliner Class 57/0 version shows the upgraded chassis design with the wiring safely confined to the chassis itself. Note the special slot in the top of the chassis frame for accommodating a decoder. The ESU sound decoder will also fit.*

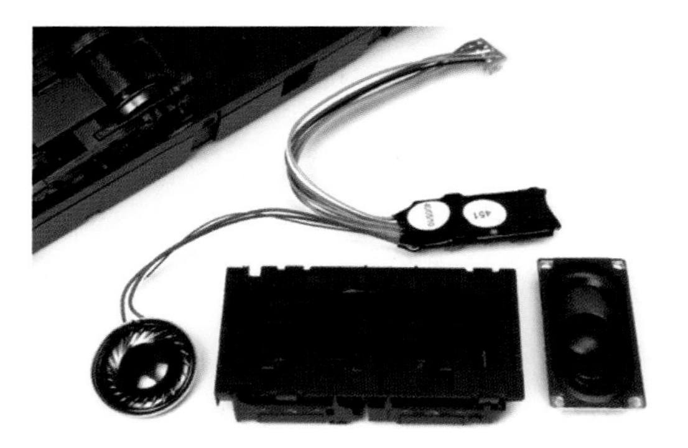

*The obvious place for a speaker is in the fuel tank/ battery box moulding, which is clipped to the underside of the chassis.*

*A hole is drilled through into the main body space to allow the speaker wires to be routed between the decoder and the underframe detail.*

*It appears that an oblong 20×40mm speaker may fit the underframe fuel tank/battery box moulding of the Freightliner version, which does not have a lighting changeover switch fitted to the underframe.*

*A 23mm round speaker is also checked for fit.*

*The speaker wires are fed through to the main chassis area.*

*The 23mm-diameter round speaker takes up half of the underframe space on the FGW Class 57/6 version, which is just as well because the other half accommodates the lighting changeover switch. There was insufficient space for the speaker enclosure, so one was made up from scraps of styrene card.*

*The completed project at work on the author's layout showing off its running lights. The speaker located in the fuel tank managed to produce a reasonable volume even when placed in such a confined space.*

## HO SCALE GP38-2 BY PROTO-2000: DIGITAL SOUND IN AN OLD FAVOURITE

Locomotives with long, narrow bonnets are particularly challenging to equip with digital sound. The Proto-2000 HO scale model of the ubiquitous GP38-2 is an interesting project, since the chassis frame has to be partly stripped of equipment and the ballast weight modified to make room for the decoder and speakers.

*Geeps or general purpose locomotives are the maids of all work in the USA, found on switching duties, maintenance of way work and local branch freights. Every US-outline model railroad will have at least a pair like GP38-2 No. EMDX 785 and GP39M No. BNSF 2898. This project shows how working lights can be fitted to small general purpose locomotives.*

When it comes to finding a suitable decoder, US-outline modellers can choose from a variety of DCC sound decoders, some with generic EMD diesel sounds and some with specific recordings from actual locomotives, including the GP38-2. Sound recordings can be variable in quality because they are affected by the prevailing conditions when the recording was made. While tempted by the MRC sound decoders, research revealed that ESU decoders were available with specific sounds for GP38-2 locomotives, including variations in the horn and bell. The ESU decoder used in this project was purchased from Litchfield Station, together with twin speaker sets with a common 16×35mm enclosure. The speakers are wired in series for correct operation. The normal 23mm-diameter round speaker supplied with ESU decoders and the larger 40×20mm oblong speaker, together with their enclosures, would not fit in the long bonnet of the GP38-2. For the record, the ESU decoder measures 31×15.5×6.5mm and just fits the long bonnet of a GP38-2 model.

*A view of the mechanism used in the Proto-2000 model. It weighs 528g, owing to the large pieces of ballast weight that have been squeezed into the long bonnet area of the model. Its hauling power is impressive, given that it has only four driven axles.*

*There appears to be an 8-pin socket for DCC conversion, but closer examination reveals that it does not function as such and should be removed from the model. Sometimes, if you're in doubt about the integrity of electrical circuits in a model, remove them and solder the decoder harness wires directly to the motor and pickup terminals.*

*A close look at the decoder and the standard speaker supplied with it. Note how much more practical the narrow twin-speaker set is for this application.*

*The factory-fitted circuit board and various sockets are removed from the model and discarded.*

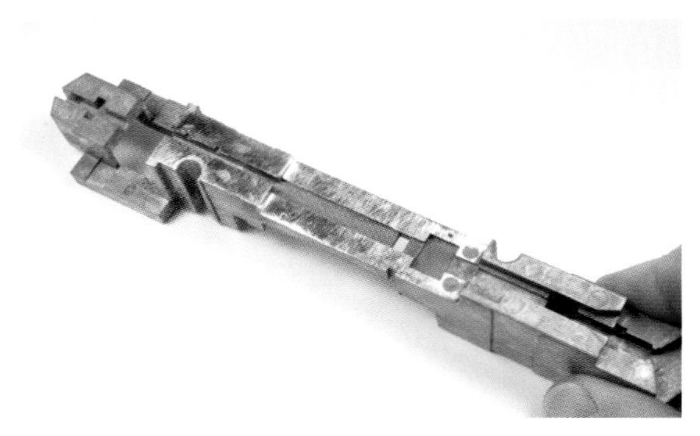

*The interior ballast weight is easy to remove by releasing screws located in the underside of the chassis frame. About 3mm of metal is filed away from the top surface of the ballast weight to create more space for the speakers and decoder.*

*LEFT: ESU sound decoders programmed with GP38-2 sounds by Litchfield Station of Avondale, Arizona. The decoders are supplied with a standard 23mm-diameter round speaker that is too large to fit GP38-2 models. There is the option to buy a smaller twin speaker set, measuring 16×32mm, including the enclosure, that does fit the long bonnet of the model.*

*Experiment to determine the optimum position of the various components. Insulation tape is applied to the metal surfaces to which components would be attached, to prevent accidental contact.*

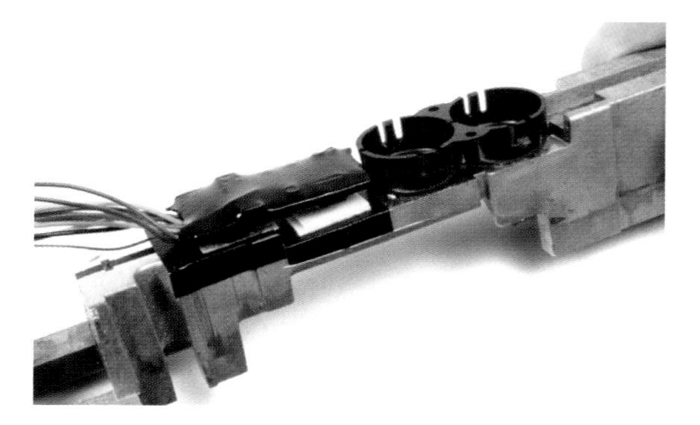

*It is surprising how much space the harness wires and the cables can take up within the interior of a model. Try to reduce the amount of cabling to the absolute minimum, which will make reassembly of the model much simpler at the end of the project.*

*It's coming along nicely, with the ballast weight refitted to the chassis frame and the various harness wires being soldered to terminals made of pieces of strip board.*

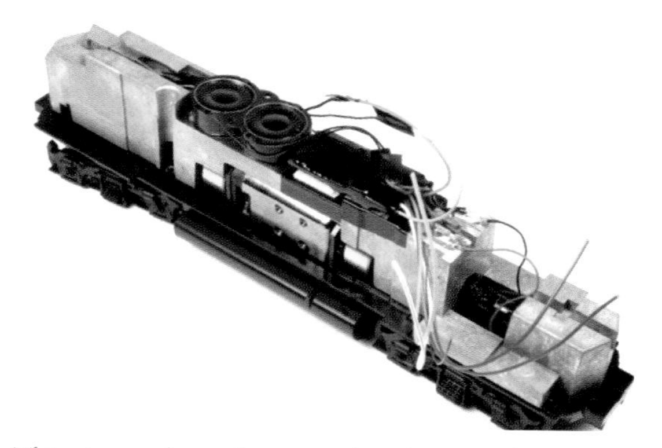

*This picture shows the connections between the motor, track supply and decoder soldered in place. The harness wires relating to output functions to operate the running lights have not yet been connected (white, yellow, green, mauve and blue harness wires) and this will be done when the model is reassembled. When replacing the locomotive body, it is important to ensure that no pressure is applied to the decoder from the inside of the bodyshell that might damage delicate components.*

## VITRAINS CLASS 37: AS SIMPLE AS THEY COME

English Electric Type 3, Class 37 locomotives have a distinctive beat, which was once commonly heard all over the British rail network. While the full-size Class 37s have dwindled in numbers, modellers can re-create that distinctive sound using a digital sound decoder. Bachmann released its own version of digital sound in a small selection of its Class 37 models in late 2007, but many of their competitors remain 'silent'. Modellers keen to equip their 'silent' DCC models with sound can use an ESU decoder and speaker programmed with bespoke sounds recorded and processed by Steve Weeks of South West Digital. The ViTrains Class 37/4 model was chosen as the host model for this installation project.

*Two EWS Class 37/4 locomotives idle at the head of a heavy charter train at Nairn in August 2007. Sound decoders enable you to re-create the sounds of such locomotives. Even when they are not on the move, the models will 'tick over' at idle until you power them up with the throttle. Random sounds recorded into the decoder's memories can include those heard from the full-size locomotives, for example from compressors, cooling fan motors and air dryer equipment.*

*New to the British market is Italian model train manufacturer ViTrains. During 2007 it released new models of the ever-popular Class 37, including eight versions of the Class 37/4 and several Class 37/0s, complete with a smooth-running and reliable chassis.*

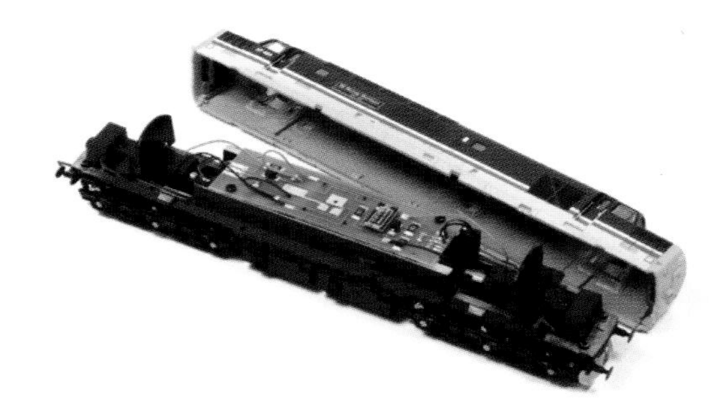

*The model weighs in at more than 450g, but still has room for a decoder among all the die-cast metal and internal circuit board. There's room for a speaker, too.*

*One problem with some decoders is that the harness leads are not sufficiently long for the decoder to be placed in a safe location in the model and still reach the NEM 652 interface socket. There's no such problem with the ViTrains Class 37 chassis.*

*Speakers have to be located in a secure place, too, so they do not work loose during use. They also must be fitted with their enclosures for optimum sound reproduction, which takes up a surprising amount of room. There is space over the mechanism, but that must be prepared to accept the speaker.*

*A piece of 40thou styrene is installed to form a fitting platform over the drive mechanism.*

*All is found to fit satisfactorily, without the speaker enclosure touching the bodyshell interior and causing unwanted resonance. The speaker and decoder are fitted next and then tested on the service track.*

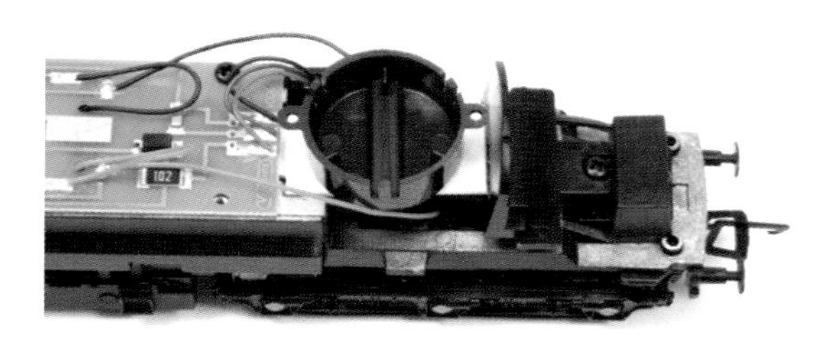

*The empty speaker enclosure is fitted temporarily with double-sided adhesive tape and the body refitted to check that it clears the internal surfaces, particularly the curvature of the roof at the shoulders of the locomotive.*

## HORNBY CLASS 60: TOO MUCH INTERIOR DETAIL FOR COMFORT?

Hornby has done a lovely job of modelling interior details in its Class 60 model. That detail is visible through etched metal grilles in the side of the model, detail that is worth enhancing with digital sound. The same detail, however, takes up space that could be occupied by a decoder and speaker. Further investigation revealed that the underframe tanks also have a lump of ballast weight concealed inside the moulding. So how is a Hornby Class 60 to be fitted with the ultimate in detailing?

*A side view of the speaker installation.*

*Class 60s were built by Brush and entered traffic at the beginning of the 1990s. They have distinctive sound that is available as a bespoke sound project programmed in ESU decoders.*

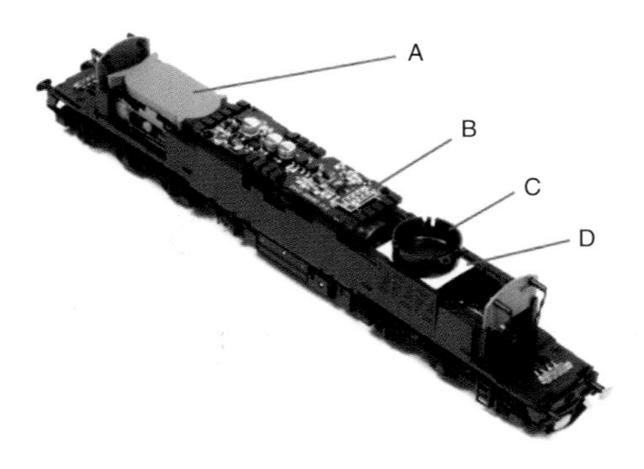


*Much of the interior space is crammed with detail at the cooling group end of the model (A) and a large complex circuit board with NEM 652 interface socket (B). A speaker enclosure (C) is placed at the No.2 end to show how tight things are with this model. Removing the lovely detail at the No.1 end is not an option! Note the use of a piece of styrene card to provide a safe place to locate the speaker (D). The decoder could sit on top of the detail at (A) if the speaker leads are extended to suit.*

*The only way to fit the standard 23-mm diameter round speaker is to reduce the size of the speaker enclosure.*

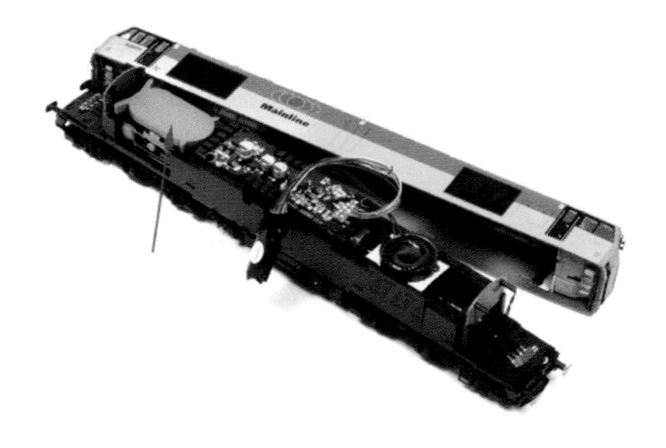

*The modified enclosure still produces good volume but does not come into contact with the locomotive bodyshell after reassembly. The decoder fits on top of the interior detail indicated with the arrow.*

## HORNBY CLASS 31: A CREATIVE SOLUTION TO FINDING SPACE

There are times when sacrificing a piece of equipment is a small price to pay to find space for a sound speaker. In the case of the Hornby Class 31, many DCC enthusiasts discard the internal mechanism that drives the cooling fan – something very much regarded as a gimmick!

*The Hornby Class 31 barely has sufficient room for a conventional decoder unless a deck of styrene is built into the chassis, located under the circuit board. To fit a speaker is even more challenging unless something is sacrificed.*

*A mechanism that drives a rotating cooling fan takes up much valuable space. Given that many modellers disconnect the fan drive because it affects the performance of the model, it's superfluous to requirements. All that is required to discard this is to release the screws.*

*Speaker enclosures can be used to check if there is sufficient space in the model. A deck of 20thou styrene has been glued to the chassis frame where the fan mechanism used to be located. A 40×20mm speaker enclosure is test-fitted using double-sided adhesive tape.*

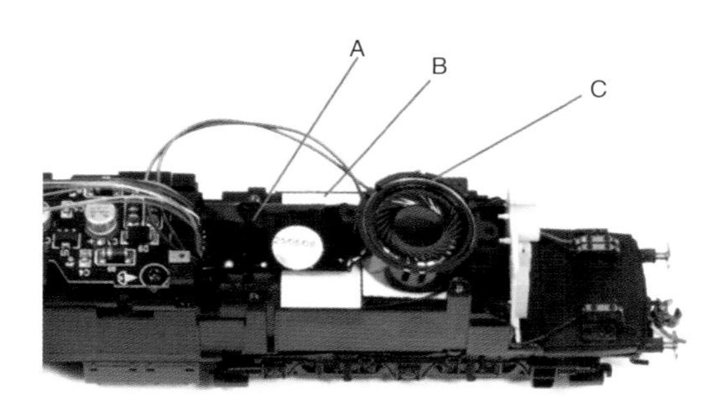


*The decoder (A) can be accommodated alongside a 23mm-diameter round speaker (C), which appears to be a better fit altogether. Both speaker and decoder are located on a floor of 20thou styrene to prevent contact with the bogie gear tower (B).*

*Reassembly is straightforward with no part of the body coming into contact with decoder or speaker. Some work to restore the fan as a cosmetic detail is worth considering so the internal workings of the model are not visible through the fan grille.*

CHAPTER 7

# Setting up for Layout Operation: Train Driving on Your DCC Layout

*DCC puts you in the cab, just like in this Class 47. Your hand throttle (cab) gives you all the controls to drive a train, including forward, reverse, braking, lights, throttle (power handle), speed measurement (in speed steps) and sounds such as horns. Your throttle is not a conventional model railway controller but a set of driving controls. When you get used to that concept, it changes your whole approach to layout operation.*

## AN INTRODUCTION TO DRIVING

After all that wiring, selecting decoders, plugging things together and selecting addresses for your locomotive fleet, the time has come to enjoy what DCC is all about: driving trains – just like the real thing. By this time you will have become familiar with the equipment. A few hours spent playing with the command station, a couple of locomotives, the throttle and a length of track will have been enough to show how exciting the operation of a layout will be. Now is the time to gain those bangs for the capital outlay.

In reality it's not that straightforward, because

a great deal of experience of DCC operation will have been gained while testing a layout's wiring during its construction. Even experienced users will benefit from this. However, nothing beats your first full-blown operating session, especially if it's a shakedown session with friends, a 'golden spike ceremony' or while preparing a portable layout for its first train show. Then it's time to play with all those features I have been enthusing about over the last hundred pages or so!

*The controls of steam locomotives look very different to a diesel-electric locomotive's, but the principle is still the same. Forward and reverse controls, a regulator and other controls still relate to those on your throttle.*

*ABOVE: Another locomotive cab. This is the control stand in a MRL SD45-2 and again the controls are similar: train brake, dynamic brake (in the case of this mountain machine), locomotive brake, reversing lever and throttle. The switches control locomotive systems such as lights, horn, bell, sanding gear and safety devices. The instruction notices relating to each control are an interesting feature.*

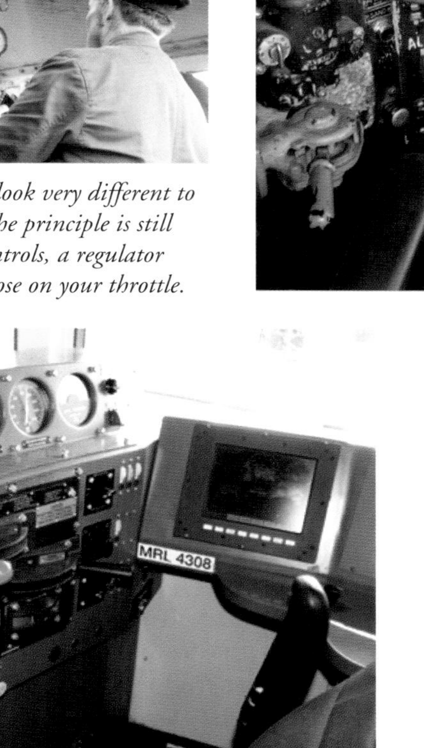


*The roomy, comfortable cab of MRL 4308, a 2006-built EMD SD70ACe. The environment is cleaner, smoother and more sophisticated than that of the SD45. None the less, the main controls remain the same and can be directly related to those on the DCC throttle.*

*Some US railroads use remote switching operations where the locomotive engineer works from ground level. The device used to control yard operations in Missoula is called a Canac Belt Pack. It's as close to a wireless DCC throttle as I have ever seen on full-size railways – even the controls are the same!*

*The complete remote switching set-up: engineer with throttle controls and a panel for setting the turnouts.*

*Look familiar? The turnout control panel has similarities with model railway practice.*

The wide variety of designs and approaches to the driving controls available with different systems will startle even the most experienced modelling enthusiast. Compared to analogue control, most DCC controls are located on the throttle, including turnout and accessory control. The choice of controls is very personal. Some modellers will choose a base station with driving controls, such as the ZTC-505 or ZTC-511, together with a hand throttle or two. Alternatively, when layout operations are centred around roaming with the train as it is driven, a DCC system based solely on hand throttles, such as the Digitrax Super Chief, Lenz Set 90 or Set 100, or the MRC/Gaugemaster Prodigy Advance, may be preferred over those using a console. Coupled to that is the added dimension of wireless operation – not being tethered to the layout at all. Bachmann offers complete DCC systems for wireless operation

*DCC users need to know the front of a locomotive from the back in order for forward direction on the throttle to move the train in the right direction. The situation is the same on US railroads, with the 'front' of a locomotive identified with the letter F located in the appropriate place. Strange as this may seem, because logic suggests that the short bonnet of a US diesel locomotive is logically 'The Front', some operators prefer to run locomotives long bonnet first, although this practice is increasingly rare. The letter F tells ground staff which way the locomotive will move when communicating with the engineer during switching operations. Also, the engineer knows which way the locomotive will move when the reversing lever is placed in the forward position.*

*During testing of the project layout, several locomotives were parked one after the other in the sidings, with running lights left illuminated. Not a single block switch was needed for this close operation of locomotives. It was all done with the throttle.*

under its Dynamis label, while Digitrax has infrared and wireless options in its range (always check to see if a wireless system is legal for your country of residence).

Whatever your preference – console or hand throttle, tethered or wireless – the basic functions of the throttle are the same, providing a set of driving controls and a method of programming decoders to receive a readout of values. The throttle puts a locomotive's cab controls in your hand. Compare the throttle controls with those of full-size locomotives in the accompanying photographs. No matter which side of the 'pond' you are modelling, the controls are essentially the same and some manufacturers have even designed their base stations and throttles with miniature versions of levers.

This chapter discusses the features that affect the driving experience, such as layout and locomotive performance, decoder settings and operating features. This includes the use of throttles, operating locomotives, adjusting decoder settings to refine performance, momentum effects and how to use DCC features for double-heading, consisting and speed matching.

## DRIVING CONTROLS

### The Reversing Lever.

Either presented as two buttons for forward and reverse (Lenz LH100, Digitrax DT400) or as a lever with a centre neutral position (ZTC master controllers and Digitrax Zephyr), the function is simply to control forward and reverse direction of locomotives and consists. That does not change the polarity of the track power, which remains at a constant voltage at all times. The decoder responds to the change in direction instruction and will move the locomotive forward when the reversing switch is set to 'forward' regardless of the actual position of the locomotive on the track. Ensure that decoders are correctly installed so that the designated 'front' of the locomotive leads when the controller is set to 'forward'.

### The Power Handle

Speed control is usually controlled with a rotary knob, speed step buttons or something along the lines of the regulator handle found in a typical steam locomotive. Rotary knobs may have a centre off position, be perpetual (no stop point) or have a clear stop point depending on the design. Speed is often shown on the handset display as speed steps; 0–28 or 0–128 depending on the chosen speed step setting. Some systems allow the operator to change the speed step setting for a particular locomotive by a push of a button during operation on the mainline, in the same manner as using a function button. The command is sent by the use of loco-specific 'programming on the main'.

### Brake

A couple of systems offer a brake feature that fine tunes the operation of the locomotive during deceleration. Otherwise, breaking is achieved by changing the position of the speed control knob or pressing the correct buttons to reduce the speed steps. The inertia effect programmed in most decoders will simulate the braking effort required to stop a train. That can be adjusted as required by changing the value of the appropriate CV, which can be effected when the locomotive is in operation on the mainline by using loco-specific 'programming on the main'.

### Function Buttons

Each function button can be used to control a variety of onboard systems: lights, strobes, bells and other sounds. They represent the buttons in the locomotive cab that do exactly the same thing, although modellers do not need sanding gear, windscreen wipers or radio controls!

Specific buttons will be allocated to accessory control so turnouts can be easily changed with a couple of button presses. While not directly related to the controls in a real locomotive cab, it's a feature that means the person operating the layout has the option of controlling turnouts when driving, avoiding the need to rush back to a central console of turnout switches. In effect, this could represent the work of a second man during local switching operations, although that could also be done using manual ground throws, too.

### LCD Screen

This is the information readout that covers for the speedometer and other information dials in the full-size locomotive cab. Those throttles with turnout and accessory control will also offer a readout of turnout positions for reference, together with the locomotive address, speed steps, activated functions and power consumption. As a technology called feedback (such as RailCom) becomes more universal, modellers will be able to refine operations by using DCC to simulate fuel levels, speed, throttle positions and other such operational features.

### Numeric Keypad

This is for selecting decoder addresses of the locomotives you wish to drive. Some modellers worry about forgetting their decoder addresses. A simple way is to use the loco number, which, in the case of British steam locomotives, US-outline diesels and pre-TOPS BR diesels, is very easy to remember because they are four-figure numbers taken from 1 to 9999. Simply check the locomotive number on the cab side, smoke box door or light boards to key in the correct address. TOPS (Total Operations Processing System) numbers are five-figure numbers for locomotives, which leaves the option of using the last four figures: Class 66 No.66 235, for example, becomes address 6235. Multiple units with six-figure TOPS numbers are treated the same way: DEMU Class 207 No.207203 becomes address 7203.

*Wireless operation is an interesting feature to build into a layout. The Lenz cordless telephone system (XPA adapter) can be plugged into XpressNet very easily using standard cables and RJ connectors supplied with the equipment. A flat stable surface, however, is required to locate the telephone base station. Other wireless systems, such as Bachmann Dynamis, require clear lines of transmission for its infrared controllers, so this technology has to be carefully considered when installing in any layout.*

## ROAMING AROUND THE LAYOUT – THE ULTIMATE IN DRIVING

While the way in which a layout is operated depends on its size, format and set-up, most modellers opt for roaming operation, especially with large layouts. The reason for this is simple: following the train around the layout gives a sense of distance and purpose, with the operator taking the role of driver or engineer, obeying signals and track conditions together with train operating instructions to get the job done – it's very enjoyable indeed.

Hand throttles can be unplugged from the throttle bus network so the modeller can move from one layout location to another, keeping up with the train. The train will continue to run even after the throttle has been disconnected because the throttle does not directly control the power in the track. The command station constantly transmits the last instruction received from the throttle and will continue to do so until the throttle is plugged back into the system and the controls further adjusted. When roaming, give yourself enough time to plug back into the system by adjusting the speed of your train. Searching for the throttle connection port in a hurry is not particularly enjoyable, especially when your train is heading for disaster!

In the event of an approaching disaster, use the emergency stop button to bring everything to a stand. There are numerous ways of doing this, depending on your chosen system, and the manuals should be read to see how it works. Some throttles allow you to bring your train to a stand simply by pressing both the forward and reverse buttons without affecting any other train. Some systems simply stop all the trains from moving by pressing the emergency stop button without cutting the track power. This enables turnouts and other accessories to be changed when trains are at a stand so misaligned turnouts can be corrected. Finally, there is an absolute stop position, which cuts all power to the layout when you have to cut everything to protect electronic devices. The stop button may also be used to reset the system once the problem has been resolved. Sometimes a reset button must be activated instead.

Wireless systems are becoming more common and they truly enhance the enjoyment of roaming operation. No longer do throttle cables become tangled at busy locations and there's no need to find a throttle port to stay in control of the train. When setting up wireless or infrared throttles, ensure that there are clear lines of sight between throttles and receivers. Also, be aware that some wireless systems sold in the USA are not passed for use in Europe. This is a pity because the Easy-DCC universal wireless system, the CVP Products RX900 Receiver together with the RF1300 throttle, is designed to work with a number of systems including Lenz, System One and North Coast Engineering. Unfortunately it only complies with the US regulations – not those for Europe.

This does not mean that wireless roaming is unavailable to British and European DCC users. If expense is no object, a universal wireless

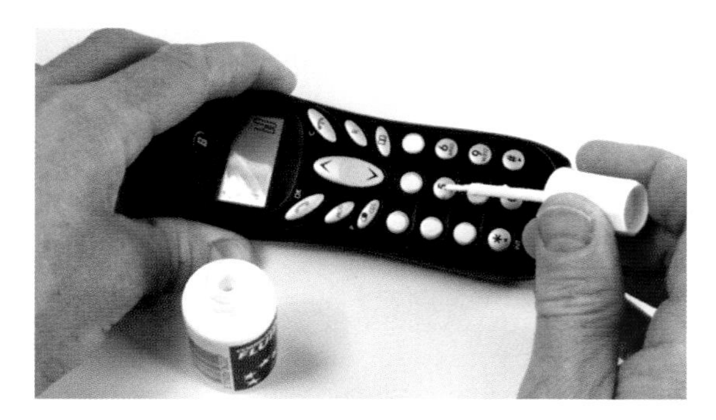

*Paint over the keys on the cordless phone handset with typewriter correction fluid or acrylic paint. A black marker pen is used to write in the new key functions.*

*The completed phone and base station with keys marked for reversing, throttle controls and function control.*

throttle and base station has been developed by ESU (Electronik Solutions Ulm) and is available for both European and US modellers. At the other end of the spectrum, Lenz offers an inexpensive adapter that converts the signals from cordless telephones into XpressNet communication. The telephone handset becomes a wireless throttle and the base station is connected to the adapter. This system has the advantage of built-in recharging of the handset batteries and using legal wireless frequencies already allocated to cordless phone operation.

Digitrax offers both a wireless system and one that operates with infrared technology that is available in Europe and the USA. MRC added a wireless system to its Prodigy Advance system, which is currently passed for use in the USA. Finally, Bachmann released its Dynamis wireless DCC system in late 2007 for the European and US markets.

During operations with wireless equipment, it is essential to have a supply of fresh batteries to hand, and ensuring that rechargeable batteries are fully charged before friends arrive for an operating session is just one of the jobs one has to do to prepare the layout. Unfortunately, most wireless DCC throttles do not plug into a base station for convenient recharging, unlike the cordless phone handset. Have a container to hand to accept spent batteries so they do not end up littered across the layout, becoming a potential hazard to the system itself.

## PROGRAMMING DECODER CONFIGURATION VARIABLES (CVs) TO REFINE OPERATIONS

As locomotives are operated on the layout, the need to fine tune performance usually comes to the fore. You may have two engines in a consist that do not work together particularly well. Some may shoot off at speed and stop suddenly with only small adjustments to the throttle. You may notice that lighting features do not perform in a realistic manner. How can performance be improved or changed to be more authentic? By changing settings in the decoder.

Programming is the feature that enables changes to be made to the way that a decoder is set up for operation, including changes to function button allocation called 'function mapping'. Programming is done by changing settings in Configuration Variables (CVs), which are, in effect, numbered pigeon holes or slots that are assigned a value. Each slot controls a different function of the decoder, whether the lighting system, addresses, inertia or some other feature.

Every decoder is supplied with default settings for each CV, one of which is a mandatory setting as defined in the NMRA Standards and RPs: CV1 Basic Locomotive Address of 03. Most newcomers to DCC get the hang of CVs pretty quickly when experimenting with locomotives and testing the layout during construction. There will have been a need to change CV1 to enter a unique address for each locomotive on the layout so they can be operated independently of each other, and in most cases modellers rarely do anything more than that. Programming can be done in one of two ways: 'Programming on the Main' and 'Service Mode' programming.

## PROGRAMMING ON THE MAIN

Programming on the main is also referred to as operations mode programming. This method changes decoder CV settings when the locomotive is in operation on the layout and can be used to set momentum effects and even turn digital sound volume up and down. Low-cost entry-level sets offer only programming on the main as a way of programming decoders. One of the catches is that the programming instruction is broadcast through the layout and can be picked up by all locomotives present on the layout at that given time. This is clearly problematic. Also, this method does not protect newly converted locomotives from full track power when being tested and allocated their unique addresses. The broadcast method of programming on the main can be done on a separate length of track, using a double pole, double throw switch to switch between the layout and the separate piece of track. Further protection is offered by installing a resistor of about 1,000 ohms, which will protect the decoder if the installation is faulty.

Locomotive-specific programming on the main is a feature of advanced systems. For this the address of the locomotive to receive the programming instruction has to be entered on the throttle. Only that decoder will act on the programming instruction, all others will ignore it. One drawback is that the old and new CV value cannot be read on the throttle display unless the layout has feedback circuits built into it and the decoder supports feedback. The decoder address cannot be changed either, because the address is needed to identify the decoder being programmed.

## SERVICE MODE PROGRAMMING

Service mode programming is carried out with an independent service track and a dedicated output from the command station. The power is much reduced, protecting the decoder against faults in the locomotive while values are read, the installation tested and the locomotive allocated a unique address. Unlike programming on the main, it is not necessary to know the decoder address and it can be used as a method for confirming an address if it has been forgotten.

The decoder installation is always tested first, by checking the readout for CV1, which should be default address 0003. An error code will be shown instead of the default address. When this happens the installation should be checked. Any further changes to CVs should only be done after the installation has been successfully checked.

Programming of decoders is not as onerous as it sounds, since it only requires a change or changes to predetermined slots within set criteria. The decoder manuals offer a great deal of information on the value of each CV and what it does. All the modeller has to do is change the CV values according to the instructions. There are several ways in which DCC systems allow you to do this in service mode:

- 'Direct' mode offers the simplest method for changing settings in key CVs such as the address, acceleration delay, braking delay and other similar operating CVs. It saves time by offering a specific button or term in the menu. Changing CVs using this method is quick and easy, nor do you need to remember the CV number. The DCC throttle manual will explain the options for direct programming.
- 'Paged' mode is used to programme decoders that do not support the direct model. It's regarded as slower but is favoured by some manufacturers. None the less, it still enables the operator to directly change the value of CVs.
- 'Register' mode is an increasingly obsolete method of programming older decoders that do not support the direct or CV mode.

Some CVs are changed by 'bit switching', a process that has many newcomers to DCC breaking into a sweat but is actually very easy. It avoids a lot of complex mucking around with HEX as well. Of all the CVs in a decoder, CV29 is the key settings' CV, which governs some of the most important features such as analogue operation, speed steps and the use of extended addressing so that four-figure addresses can be used. While I rarely need to change anything in CV29, if necessary it is done by selecting the required setting and entering that as a CV value in CV29. For example, value 3 is for 'Operating' mode, which can switch the decoder from DCC only operation to DCC and analogue operation. It's a simple 0 and 1 switch – a bit value of 0 switches analogue capability off and a bit value of 1 switches it back on.

Other CVs use bit switches to switch features on and off as well, such as BEMF, constant braking features and anything else the manufacturer has built into the decoder.

There are a handful of CVs that are mandatory and some that are recommended in the NMRA Standards and RPs. Some can be changed by the modeller and some are fixed, such as CV7 and CV8.

| | | |
|---|---|---|
| CV1 | Mandatory CV | Primary Address: default value is 3. |
| CV2 | Recommended CV | V-start or minimum starting voltage. |
| CV3 | Recommended CV | Acceleration rate: determines the starting momentum of the decoder. |
| CV4 | Recommended CV | Deceleration rate: determines the braking momentum of the decoder. |
| CV7 | Mandatory CV | Manufacturer version number: reserved for the manufacturer to store information regarding the version of the decoder. |
| CV8 | Mandatory CV | Manufacturer ID. |
| CV29 | Mandatory CV | Decoder settings. |

The allocation of standard CV numbering, with some key ones such as CV1 address being effectively cast in stone, ensures product interchangeability. The NMRA also suggests optional CVs, which are left to manufacturer discretion. Commonly found in all decoders and very useful for adjusting locomotive trim and performance are:

| | | |
|---|---|---|
| CV5 | Optional CV | Maximum speed: can be used to cap the top speed of a locomotive. |
| CV6 | Optional CV | Mid speed or V-mid: useful for further refinement to the speed curve. |

Other CVs will relate to various decoder functions including lighting effects, sound and motor drive settings. Two that you should be aware of are CV17 and CV18, which enable extended locomotive addressing. In other words, if you enter a four-figure address when programming a decoder, those CVs are kicked in automatically by the DCC system to provide a seamless address entry.

## DRIVING WITH SPEED STEPS

Should it be 14, 28 or 128 speed steps? A good question! All modern decoders will permit operation with 28 or 128 speed steps and some throttles will allow the operator to switch easily between the two to suit operating conditions; 28 speed steps allows maximum speed to be achieved in large steps, while 128 steps is ideal when fine levels of control are needed. I find that my N scale Kato and Atlas models perform best on 128 steps, whereas, when combined with inertial effects, 28 speed steps is ideal for many of my 4mm scale models. It takes trial and error to achieve the right balance. However, both the command station and decoder must support 128 speed step operation for it to work.

Few DCC enthusiasts bother with anything else, even when 14 step functionality is available to suit older decoders, many of which are not capable of 28 or 128 step operation.

There are suggestions that 128 speed step functionality really should be called 126 step, on account of the maximum value of 127 that can be encoded in the 7-bit binary field transmitted in a 128 step speed control packet, together with another factor, which really means the range is 2–127. As far as practical operations are concerned, it's an academic point and does not affect the operation of your layout. I never give it a second thought when driving my trains.

## INERTIA OR LOCOMOTIVE MOMENTUM

Inertia is the rate at which a locomotive accelerates and decelerates when the speed is adjusted on the controller. It is used to simulate the momentum of trains, especially heavy trains that need longer distances to come to a stand or take more time to get rolling. Modellers like to use momentum effects to prevent a train from coming to a stand more quickly than is considered desirable. Momentum effects are adjusted in the decoder CV settings and this can be done during operations by using programming on the main to suit a change in operating conditions, as when, for example, the locomotive has been coupled to a heavy train.

The use of inertia is a personal thing and can cause some practical operating problems on short end-to-end layouts. Some modellers prefer to have absolute control, using their skill on the throttle to simulate train handling conditions. It may also be the cause of collisions and accidents when control is effectively lost, and most modellers do not wish to have that level of operating realism!

Momentum is adjusted in CV3 (starting delay or starting momentum) and CV4 (braking delay or braking momentum), where the value 0 disables both on a range of 0–255. Some decoders have default momentum settings that can be overridden by the use of function buttons. This is a characteristic of modern Lenz decoders.

## ANALOGUE OR NOT

Most DCC systems permit a single analogue locomotive to be operated on a digital layout, using address 0. It is intended as a short-term measure and is not good practice for long-term use. On the opposite side of that particular coin, decoders can be set to work on both DCC and analogue layouts and this may be the default setting for many manufacturers' decoders. What that means is that your favourite digital locomotive can be set to work on an analogue layout, which is especially useful if your friend's layout is not digital. The setting is changed in CV29, bit 3 (operational mode) in the decoder. However, there are two important things to consider when operating a decoder-equipped model on an analogue layout: high-frequency feedback controllers may damage decoders and similar high-frequency track cleaners, normally harmless to analogue models, will be fatal to decoders.

## SPEED MATCHING LOCOMOTIVES

There are benefits in matching the speed performance of locomotives as closely as is practicable when using double-heading and consisting. Clearly, two mismatched locomotives working as a pair will not perform at their best and could lead to problems. In order to match locomotive speeds by adjusting CVs, you may either use programming on the main or operational programming, by which the adjustments can be made without having to transfer the loco to and from a service track.

The first method involves a given length of track, a stop watch and, of course, the DCC system. The locomotives to be matched with each other may have differing start voltages, different top speeds and so on. All settings for the CVs responsible for performance and the speed curve, CVs 2, 3, 4, 5 and CV6, are set to zero in all of the locomotives. Some modellers feel that the BEMF should be switched off, too. Each locomotive is checked for the speed step reading when it just begins to move. They are checked for the length of time it takes for each to travel down the test track at a given speed step. Notes taken during this process should record the fastest locomotive and the slowest at given speed steps. Also, those with the lowest and highest start speed step should be noted and compared. When programming those CVs responsible for performance, set the start point of the slowest locomotive to match the fastest and so on with the second slowest until they all start to move at the same speed step. This is achieved in CV2, minimum starting voltage. Use CV5 to peg the maximum speed of the fastest locomotive back to the maximum speed of the slowest locomotive and so on until they all run at the same speed on speed step 28 or 128. CV6 for mid speed could also be adjusted to ensure the speed curves are as closely matched as possible. A variation on this theme is to select a locomotive in your collection that performs just so, one you are really happy with. Call this your 'control' locomotive, the one that sets the standard, and match the others to it using the technique described above. A certain amount of trial and error is needed, together with patience. It's impossible to get the locomotives exactly the same as each other – a close match is the best you can hope for.

The second technique is to use a scale speedometer device, such as TrainSpeed offered by TDP & Associates, to measure scale speed using light-sensitive sensors placed between the rails at a set distance apart for any given scale. For this method, establish a minimum speed for speed step 1 and a maximum for speed step 28. You could choose full-size locomotive maximum operating speeds for authenticity: a Class 37/4 could be set for 80mph (129km/h), for example. Given that speed matching is the objective, if locomotives of different classes are involved in consisting operations it's a good idea to select one top speed figure for them all. The minimum speed for step 1 is entered in CV2 and the setting for maximum speed is entered in CV5. The locomotives are tested with the speedometer to see how close they are to the scale speed for steps 1 and 28. Adjustments are made as necessary until all the locomotives run together at a constant speed. The 'trim' is also adjusted (*see* below).

## LOCOMOTIVE 'TRIM'

Some locomotives have an annoying characteristic in that they run at markedly different speeds when run in forward or reverse directions. This can upset your carefully established speed match performance tables. Some decoders have CVs for forward and reverse trim, so that the forward and reverse performance is equalized.

## RESCUING DECODERS

I will be the first to admit to getting lost in decoder CVs when working to refine locomotive performance. If a decoder suddenly ceases to function, it is likely that there is a conflict with CV settings or an incorrect value has been entered somewhere. Rather than track down all of the settings, try rescuing the decoder by resetting it. Find out which decoder it is and look up the reset CV value in the instruction manual (you do keep the manuals, I hope!). Enter the value indicated in the manual into the appropriate CV. In the case of Lenz decoders, the value of 33 should be entered in CV8. Digitrax, ESU Loksound decoders and SoundTraxx Tsunami decoders are reset by entering the value of 8 in CV8. NCE decoders are reset in CV30 with a value of 2. Usually, this clears any problem and programming can start all over again. If that fails to work, return the decoder for repair or replacement.

*Controlling a multiple locomotive consist with a single address and matched performance is one of the benefits of DCC. US-outline modellers routinely model locomotive line-ups such as this BNSF freight photographed cresting the 'False Summit' of Marias Pass near East Glacier, Montana.*

## MULTIPLE WORKING AND CONSISTING

This is the exciting operating feature that enables double-heading of trains or the assembly of locomotive consists of three or more locomotives, all able to operate together with a single address. The operational benefits are numerous, including direct control of all the locomotives in a consist without the worries regarding performance that come with analogue control, where speed matching and trim are almost impossible to adjust. All of the programming work to adjust the speed performance and trim of locomotives comes to the fore when you start using multiple locomotive consists. There are three types of consisting mode, although not all DCC systems support them, particularly low-end entry-level systems.

### Basic Consisting

This method is as simple as programming two or more locomotives to run on a single address. All the engines must be positioned on the track to run in the same direction, which means that varying the direction that the locomotives face has to be accounted for by reversing the running direction setting in CV29 for those facing backwards. Adding and removing locos from a consist requires reprogramming of the locomotive address, which means removing them from the layout to a service track. Bachmann EZ

*Double-heading of locomotives works for steam locomotive models, too. This scene was photographed at Kingussie in April 2007.*

Command and most other basic systems rely on this method, which has the disadvantage of being inflexible. There are certain times when pairing locomotives using basic consisting makes sense, such as pairs of Class 20s, for example. If they are to run 'long bonnet facing long bonnet' rather than 'elephant' style, one will require the travel direction setting to be reversed.

## Decoder-Assisted Consisting

This is a method that stores the consist instructions in the decoder of each locomotive in the consist. The decoder retains its own address and a consist address too, which is allocated when the consist is assembled.

This method has the benefit of being flexible and locomotives are easily added and removed from the consist. Operating direction of the locomotive is not important and it can be run in either direction. The consist can be transported to another location, but extreme care must be taken to reassemble the consist in the same order as it was set up on the home layout. Also, all of the decoders in locomotives being assembled in such a consist must be capable of advanced consisting.

## Command Station-Assisted Consisting

Also known as universal consisting, this method can assemble consists with locomotives equipped with any decoder. The information regarding this type of consisting is stored in the command station and not the decoder. One requirement is that locomotives being assembled in a consist are set up to run in the same direction on the track regardless of the position of any direction indicator arrow, because direction of travel is locomotive dependent. They do not necessarily have to be coupled to each other either – the second locomotive could be located mid-train or pushing at the rear. Locomotives can be added or removed in any direction or orientation that suits operations at the time. Usually the address of the first locomotive, called the 'top locomotive address', controls the consist. There is little doubt that this is the easiest type of multiple working and some DCC systems set this type of operation up through the use of a dedicated button on the throttle rather than in a menu. It has the added benefit of being able to dial up any locomotive in the consist to operate running lights.

## Creating an Advanced Consist

The following sequence of photographs shows how two locomotives may be formed in an advanced consist using the Lenz LH100. The LH100 also sets up command station assisted consists (universal consisting). Advanced consisting stores the multiple working information in the decoder, not the command station, and that process is demonstrated in photographs using the Lenz LH100. Lenz uses simple menus to access the consist formation controls

*Two Class 37 locomotives are to be coupled together to work a heavy freight train on the project layout featured in Chapter 3. Class 37, No. 37 075 is already in position on the train and has a decoder address of 7075.*

*Another Class 37 is to be coupled to it to provide additional traction. 37 674 arrives, pausing before coupling up to 37 075. No isolating switches are used to separate the two models – DCC enables close operation like this.*

*Coupling up gently. 37 674 is the lead engine, with a decoder address of 7674.*

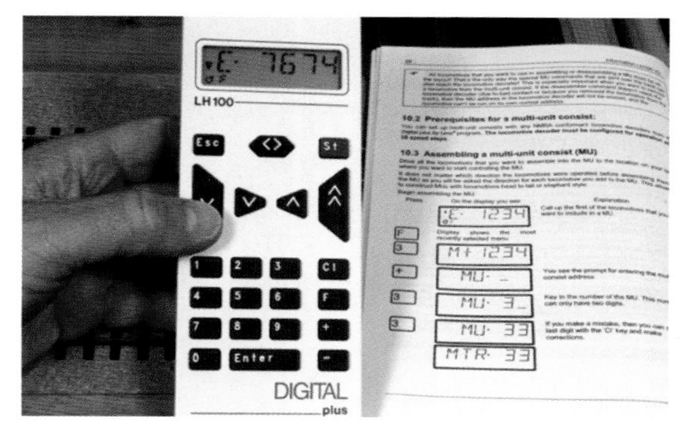


*The Lenz LH100 shows the address for 37 674 is active as an engine (the letter E is shown). It has been called up to become part of the advanced multiple traction consist.*

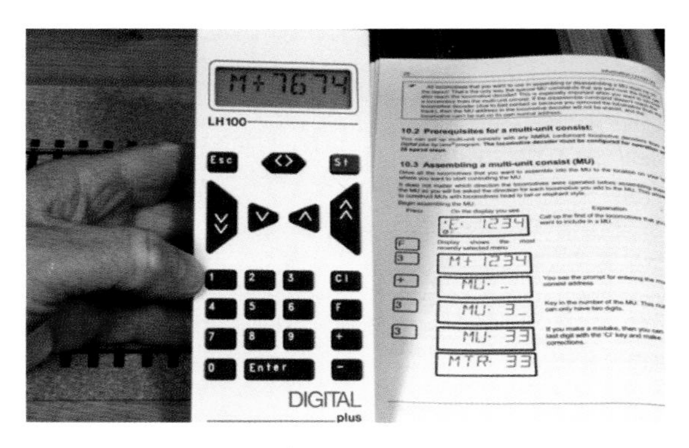


*Entering menu 3 on the Lenz system (press F and 3) shows the locomotive address being readied for addition to a consist.*

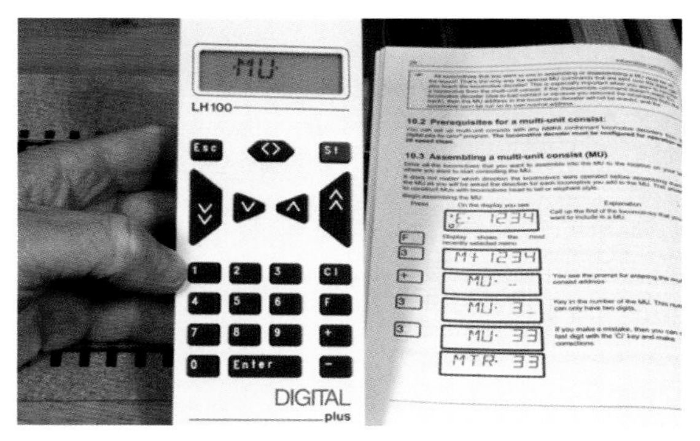


*MU is the prompt for entering the consist address. This will be the single address that the consist will respond to as if it is one locomotive.*

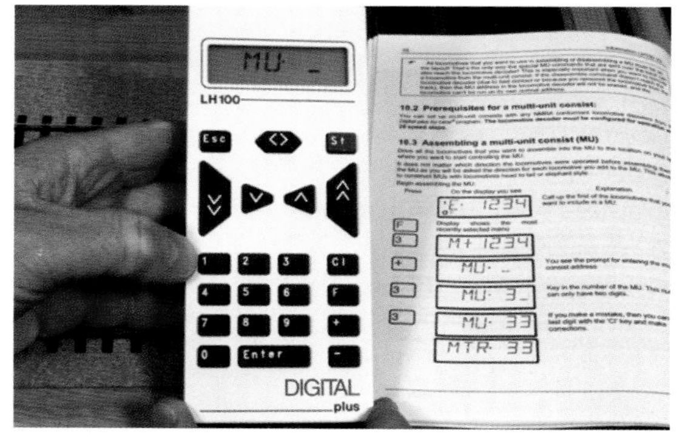


*A two-figure number should be entered.*

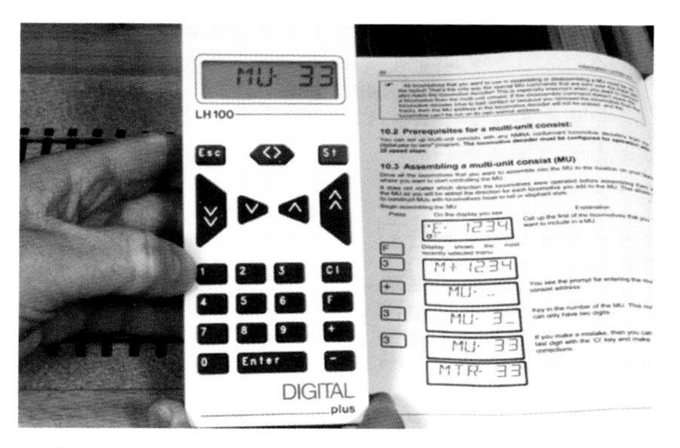


*I chose the consist address 33.*

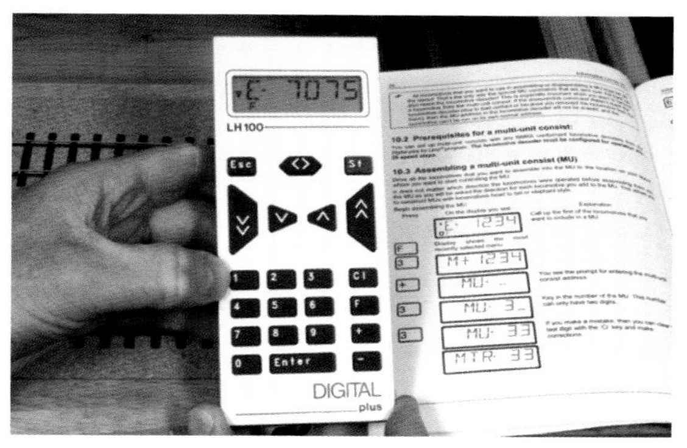


*The second locomotive, 37 075, is called up.*

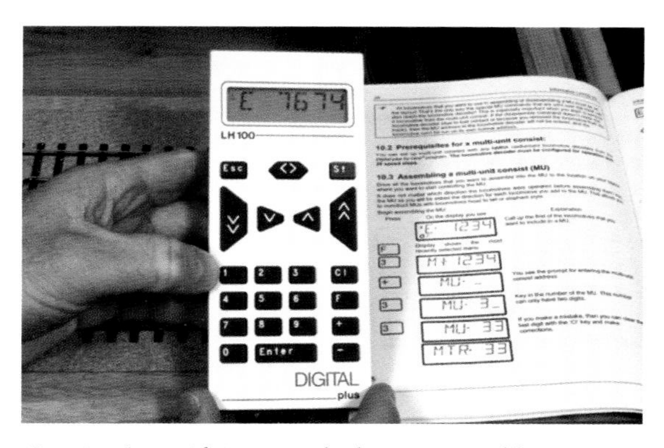


*Pressing 'enter' brings up the locomotive address again. The system prompts the operator to select the locomotive direction relative to the consist. The direction arrow, shown here, flashes.*

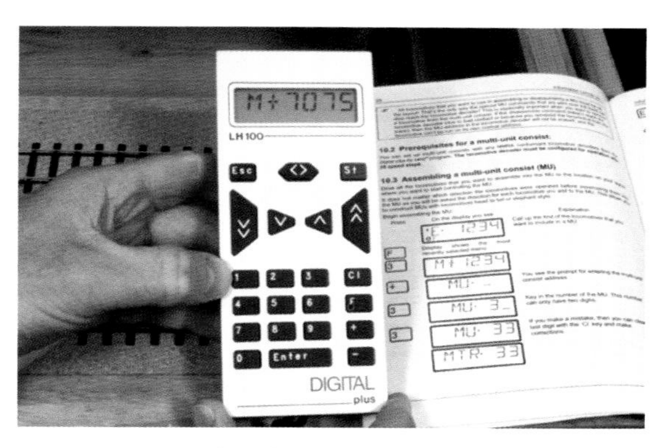


*37 075 is entered in the consist in the same way as the first Class 37.*

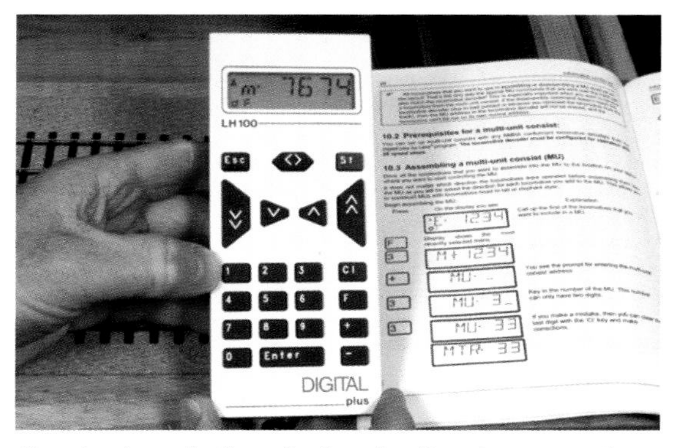


*Pressing 'enter' after selecting the direction enters the locomotive in the consist. This is indicated by the letter M before the decoder address.*

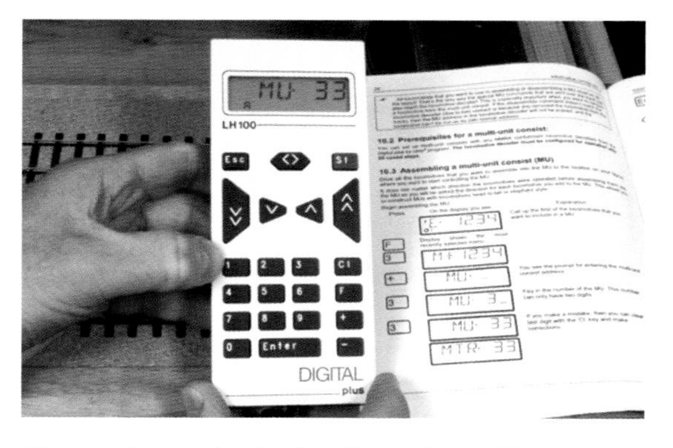


*The consist number is also allocated to 7075, ensuring that both locomotives respond to the same consist address of 33.*

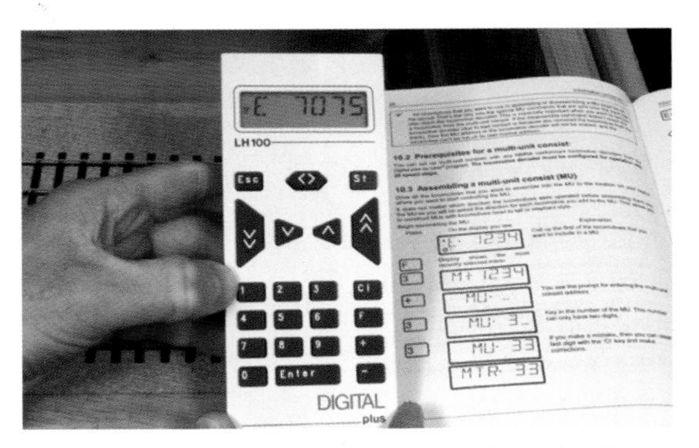


*The direction relative to the consist is chosen for the second locomotive.*

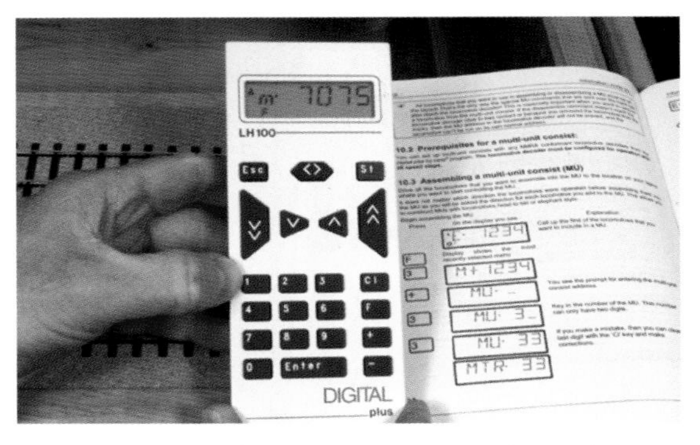


*The direction and locomotive are entered into the consist, as shown by the letter M prefixing the decoder number.*

*The consist of two locomotives is ready to go. If the second locomotive shows incorrect running lights, it is possible to call it up in the consist to operate the functions as if it is an independent locomotive.*

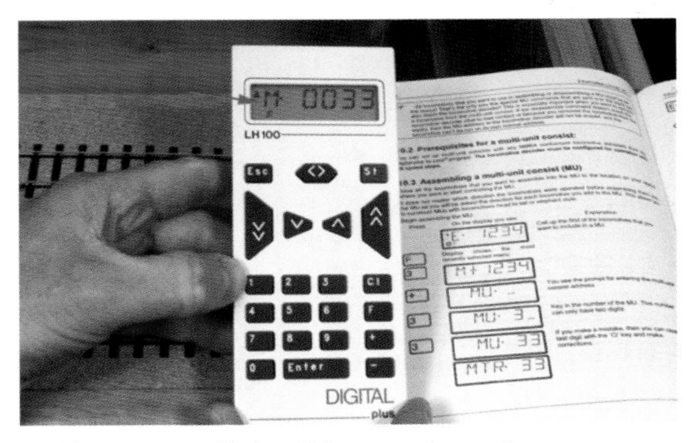


*Address 33 is called up like an ordinary locomotive decoder number and the consist driven as if it's one locomotive, not two.*

*A consist can be disbanded by simply entering the consist address. Entering the consist menu area by pressing keys F and 3 on the controller brings up a '-' symbol.*

*Hit the '-' key and the display flashes. Press 'enter' and the consist disbands automatically, leaving 'E 0033' as the display. The two locomotives are now free to work independently of each other.*

## DOUBLE-HEADING LOCOMOTIVES IN A COMMAND STATION-ASSISTED CONSIST

An alternative method used on some DCC systems is a consist function button, which some modellers may find simpler to use. The ZTC-505, for example, employs a simple process for double-heading locomotives into a command station assisted consist, which differs from the Lenz example above.

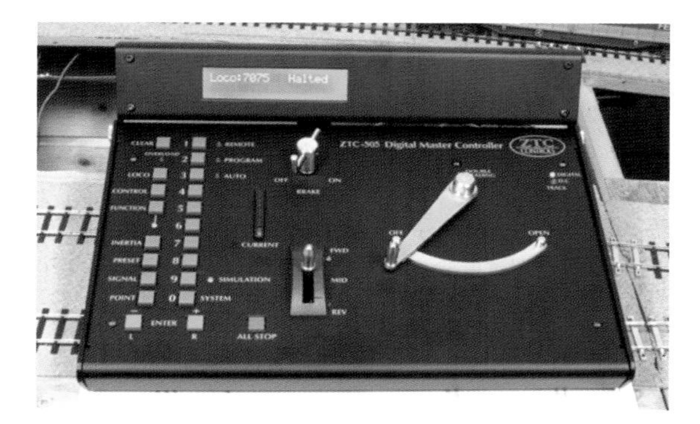


*The same pair of Class 37s is used in the formation of this consist. 37 075 (7075) was already called up on the system for when it was moved onto the head of the train.*

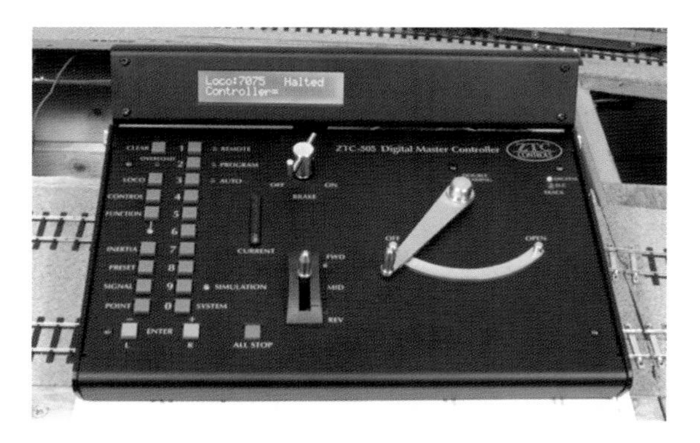


*The first locomotive to be entered in the consist, 37 674, was called up next to move it into position at the front of the train. This will be the first locomotive to be entered in the consist and will become the 'lead locomotive'.*

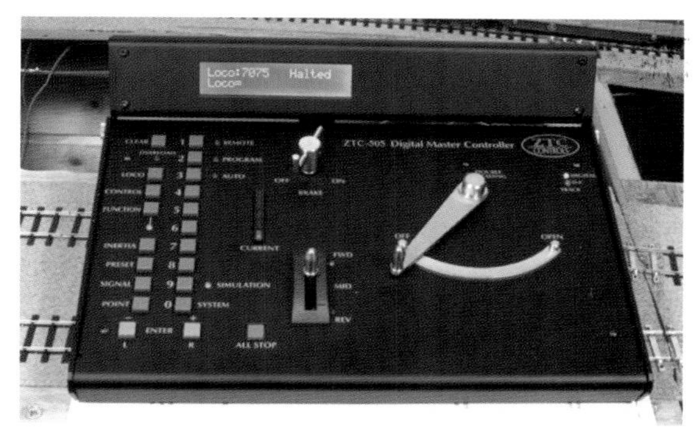


*The decoder number for 37 674 is entered using the 'Loco' command button. Up to five locomotives can be entered in this manner on the ZTC system before having to press the 'enter' key.*

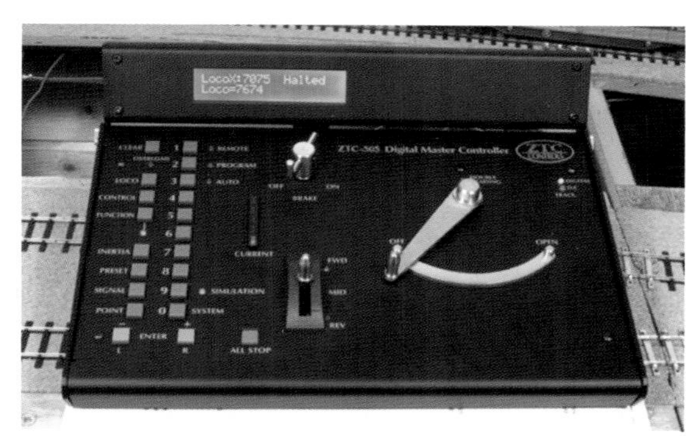


*37 674 is called up under the loco command, displacing the original call-up for 37 075.*

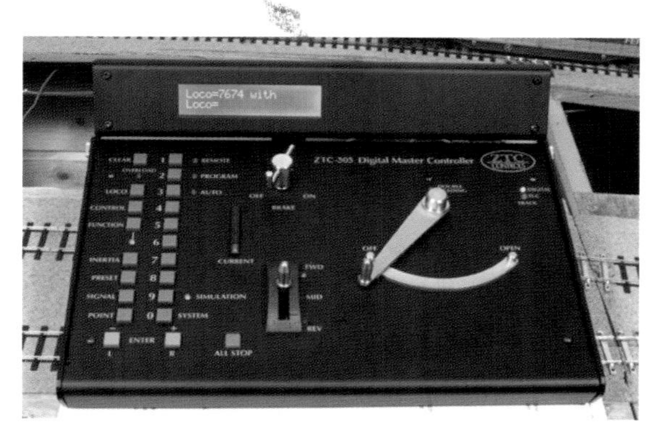


*The second locomotive is entered using the loco command button. The display shows the word 'with', indicating that these two models will be working together.*

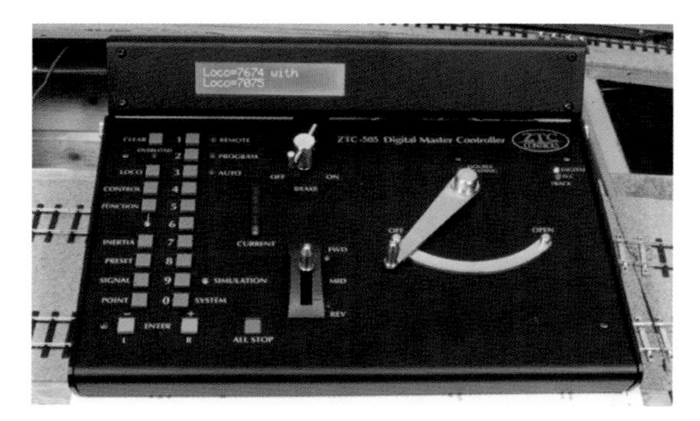


*It happens to be 37 075. The decoder addresses are entered in this order to ensure that the first address is the lead locomotive of the consist. The consist will take the lead locomotive's decoder address to control the consist.*

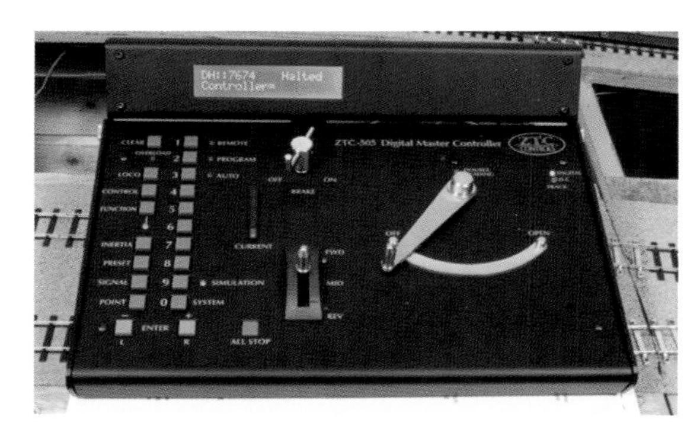


*Bringing the consist to a stand and preparing to disband it. The control button is pressed to bring up this display.*

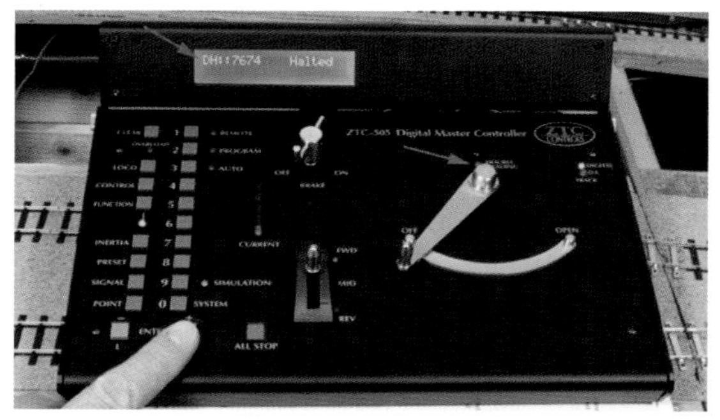


*The double-head consist is formed by pressing the 'enter R' button. This is confirmed by the letters DH prefixing the lead loco decoder address. The 'double head' LED lights up too.*

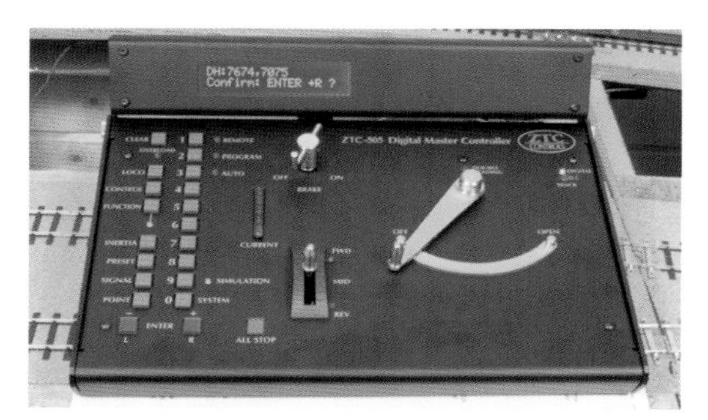


*The prompt to disband the consist is shown.*

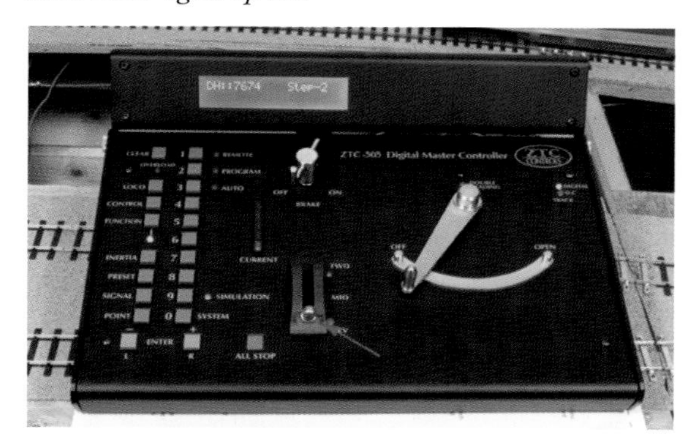


*The consist is driven using the lead locomotive's decoder address. Speed step indications and other functions are shown on the display as normal. The consist will move forward or backwards as one locomotive, depending on the position of the reversing lever.*

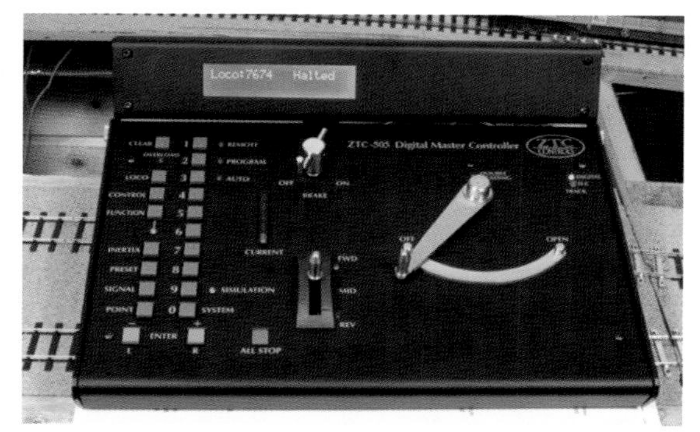


*Back to normal control.*

## IMPROVING LAYOUT PERFORMANCE

In the same manner that a decoder will not solve mechanical or electrical problems in a faulty locomotive, DCC will not improve a poorly performing layout, which can be the result of a number of factors:

- Insufficient track feed wires between the power bus and track may result in voltage drop and poor power supply. Try the coin test and install additional feed wires where a problem appears to exist (for information on the coin test and feeder wires, *see* Chapter 3).
- Do not rely on rail joiners to carry power reliably. Many faults are due to poor contact between sections of rail because a rail joiner is not making contact. Install additional feeds to bypass rail joiners.
- Nickel silver rail has high resistance to electrical conductivity compared to copper wire. Use the power bus to carry current to every part of the layout to avoid voltage drop.
- Dirty rails will prevent power from being picked up by locomotives on a digital layout – it's no different to an analogue layout in that respect. Keep the locomotive wheels and the track as clean as you can. There are numerous devices for rail cleaning: cleaning railcars, fibre pencils, track rubbers or just old-fashioned isopropyl alcohol and a lint-free rag will do the trick. Try to avoid using products that will abrade the rails excessively, since these will eventually wear away the running contact of the rail.

Whatever you use to clean the track, do not use high-frequency electronic track cleaning devices such as a Relco. They will blow decoders and interfere with the digital signal from the command station, sometimes doing serious damage. Remember, once the smoke has escaped from your base station or decoders, there's no getting it back in!

- Check the back-to-back measurements of your track. Even if out-of-gauge track does not cause a derailment, it may affect performance, especially over crossings and turnouts. Replace defective track as soon as possible.
- Check the back-to-back measurements of locomotive wheels. Out-of-gauge wheels will affect performance.
- Check that stray ballast has not become lodged between running rails and check rails on turnouts.

## FINISHING TOUCHES

*The final equipment set-up for the project layout showing throttle, command station/booster, the power supplies and a two-way circuit breaker. With the DCC project complete, work can start on ballasting, scenery and structures.*

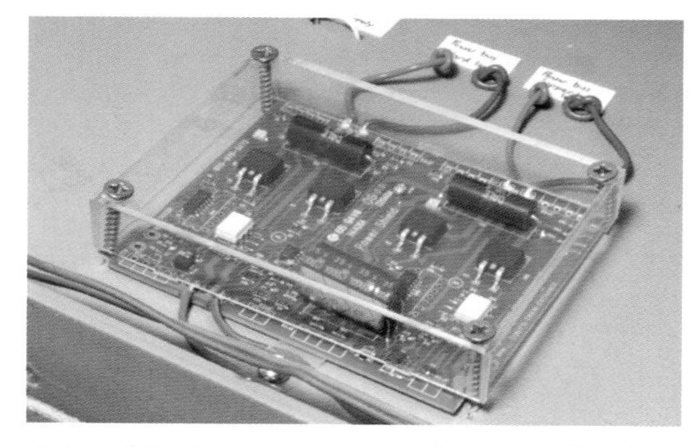

*It is useful to locate equipment such as circuit breakers in an easily accessible place, for example on the top of the fiddle yard baseboard adjacent to the control equipment. Although this exposes it to possible knocks, it is simple to protect it with a clear plastic cover. Part of a Lenz decoder box makes an ideal cover.*

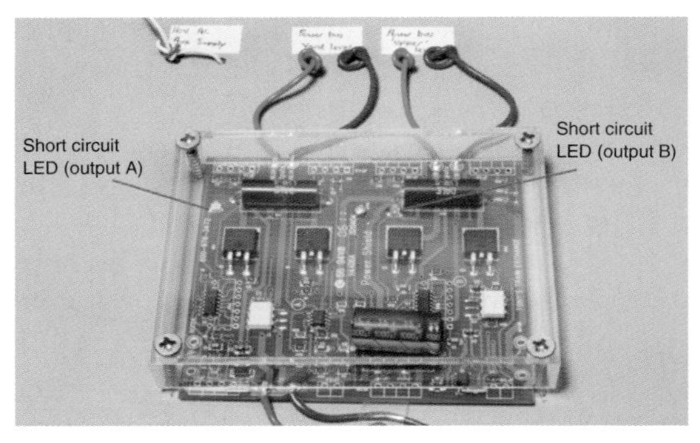


*Red LEDs on the Power Shield PS2 circuit breaker indicate if there is a short circuit, a useful operating device for fault detection.*

*Layout operation is fun and easy with DCC – no more section switches and fading lights to contend with. Both these Class 66 locomotives are equipped with sound and will sit on the layout 'yinging' away merrily until I get fed up with the noise and shut them down – with a press of key F1.*

*Speed matching locomotives is beneficial if you plan to work them in consists on a regular basis. A scale speedometer is a useful device for correlating scale speed with speed steps on the throttle so that performance can be matched as closely as possible. This device is a TrainSpeed speedometer, offered by TDP & Associates, which is designed to be built into a layout for speed checking at key locations or into a service track.*

*Test your shunting skills by running your favourite shunter or switcher on 128 speed steps.*

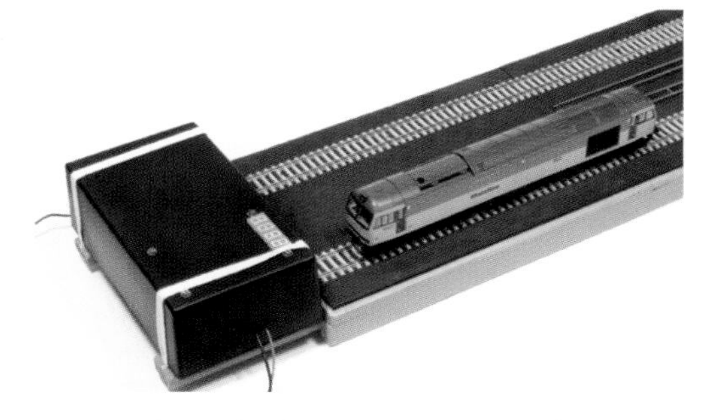

*A TrainSpeed unit built into an equipment box as part of a service track and speed testing rig.*

*Modern multiple units look really good with operating running lights and door-open indicator lights. All lighting is independently controlled from the throttle by pressing the appropriate function button, as if throwing a switch in the cab.*

*A group of stationary locomotives with lights illuminated demonstrating the 'drive anywhere, park anywhere' benefit of DCC. To create a stabling area suitable for four locomotives on an analogue layout, a minimum of four block section switches would be required, together with wiring back to a control panel. With DCC there's none at all!*

# Useful Addresses

## MANUFACTURERS FEATURED IN THE BOOK

*Bachmann Europe Plc*
Moat Way
Barwell
Leics
LE9 8EY
www.bachmann.co.uk

*DCC Specialties*
57 River Rd
Suite 1023
Essex Junction
VT 05452
USA
www.dccspecialties.com

*Digitrax Inc*
450 Cemetery Street
Suite #206
Norcross
Georgia 30071
USA
www.digitrax.com

*Gaugemaster*
Gaugemaster House
Ford Road
Arundel
West Sussex
BN18 0BN
Tel: 01903 884377
www.gaugemaster.co.uk

*Hornby Hobbies*
Westwood Industrial Estate
Margate
Kent
CT9 4JX
www.hornby.com

*Lenz Elektronik GmbH*
Hüttenbergstraße 29
D-35398 Giessen
Germany
Tel: 49 6403 900 10
www.digital-plus.de/e/index.php

*ZTC Controls Ltd*
Bancombe Road
Somerton Business Park
Somerton
Somerset
TA11 6SB
Tel: 0870 241 8730
www.ztccontrols.com

*Model Rectifier Corporation (MRC)*
80 Newfield Ave
Edison
New Jersey 08837
USA
Tel: 001 732 225 2100
www.modelrectifier.com

## VENDORS AND DISTRIBUTORS

**Lichfield Station**
*Vendor of a wide range of digital equipment including sound decoders with US locomotive sounds*

*Litchfield Station LLC*
1412 N. Central Ave Ste D
Avondale
AZ 85323-1316
USA
Tel: 001 623 298 7355
www.litchfieldstation.com/lobby/index.htm

**MacKay Models**
*Lenz distributors and stockists of Lenz and Roco equipment*

*MacKay Models*
Studio 56
Embroidery Mill
Seedhill
Paisley
PA1 1TJ
Tel: 0141 887 9766
www.mackaymodels.co.uk

**Sunningwell Command Control Ltd**
*Digitrax equipment vendors and distributors*

*Sunningwell Command Control Ltd*
PO Box 381
Abingdon
OX13 6YB
Tel: 01865 730455
www.scc4dcc.co.uk

**DCC Supplies**
*A 'DCC Boutique' supplying a variety of equipment*

*DCC Supplies*
Unit 5b
Top Barn Business Centre
Worcester Road
Holt Heath
Worcestershire
WR6 6NH
Tel: 0845 224 1601 / 01905 621 999
www.dccsupplies.com

**Digitrains**
*A 'DCC Boutique' which supplies DCC Specialties equipment and other components*

*Digitrains Ltd*
The Stables
Digby Manor
North St
Digby
Lincoln
LN4 3LY
Tel: 01526 328633
www.digitrains.co.uk

**Express Models**
*Supplier of lighting equipment including kits, LEDs and other important components*

*Express Models*
65 Conway Drive
Shepshed
Loughborough
Leicestershire. LE12 9PP
Tel: 01509 829008
www.expressmodels.co.uk

**Southwest Digital**
*Digital sound specialists and ESU distributors for the UK*

*Southwest Digital*
1 Savernake Road
Weston-Super-Mare
North Somerset
BS22 9HQ
Tel: 01934 517303
www.southwestdigital.co.uk

**Howes Models**
*Digital sound specialists and DCC equipment stockists*

*Howes Models*
12 Banbury Road
Kidlington
Oxon
OX5 2BT
Tel: 01865 848000
www.howesmodels.co.uk

**MG Sharp Models**
*DCC stockist and distributor of TCS decoders*

*MG Sharp Models*
712 Attercliffe Road
Sheffield
South Yorkshire
S9 3RP
Tel: 0114 244 0851
www.mgsharp.com

# Index